TROUBLESHOOTING ON MICROPROCESSOR BASED SYSTEMS

OTHER TITLES IN THE SERIES

NOTICE TO READERS

Dear Reader

An Invitation to Publish in and Recommend the Placing of a Standing Order to Volumes Published in this Valuable Series.

If your library is not already a standing/continuation order customer to this series, may we recommend that you place a standing/continuation order to receive immediately upon publication all new volumes. Should you find that these volumes no longer serve your needs, your order can be cancelled at any time without notice.

The Editors and the Publisher will be glad to receive suggestions or outlines of suitable titles, reviews or symposia for editorial consideration: if found acceptable, rapid publication is guaranteed.

ROBERT MAXWELL
Publisher at Pergamon Press

TROUBLESHOOTING ON MICROPROCESSOR BASED SYSTEMS

G. B. WILLIAMS

West Glamorgan Institute of Higher Education, UK

PERGAMON PRESS

OXFORD · NEW YORK · TORONTO · SYDNEY · FRANKFURT

U.K.	Pergamon Press Ltd., Headington Hill Hall, Oxford OX3 0BW, England
U.S.A.	Pergamon Press Inc., Maxwell House, Fairview Park, Elmsford, New York 10523, U.S.A.
CANADA	Pergamon Press Canada Ltd., Suite 104, 150 Consumers Road, Willowdale, Ontario M2J 1P9, Canada
AUSTRALIA	Pergamon Press (Aust.) Pty. Ltd., P.O. Box 544, Potts Point, N.S.W. 2011, Australia
FEDERAL REPUBLIC OF GERMANY	Pergamon Press GmbH, Hammerweg 6, D-6242 Kronberg, Federal Republic of Germany
JAPAN	Pergamon Press Ltd., 8th Floor, Matsuoka Central Building, 1-7-1 Nishishinjuku, Shinjuku-ku, Tokyo 160, Japan
BRAZIL	Pergamon Editora Ltda., Rua Eça de Queiros, 346, CEP 04011, São Paulo, Brazil
PEOPLE'S REPUBLIC OF CHINA	Pergamon Press, Qianmen Hotel, Beijing, People's Republic of China

First edition 1984
Reprinted 1985, 1986

Library of Congress Cataloging in Publication Data
Williams, G. B.
Troubleshooting on microprocessor based systems.
(The Pergamon materials engineering practice series)
1. Microcomputers—Testing. 2. Debugging in computer science. I. Title. II. Series.
TK7887.W54 1984   621.3819'58'0287   83–19311

British Library Cataloguing in Publication Data
Williams, G. B.
Troubleshooting on microprocessor based systems.
(The Pergamon materials engineering practice series)
1. Microprocessors—Testing.
I. Title
621.3819'5835   TK7895.M5
ISBN 0–08–029989–X (Hardcover)
ISBN 0–08–029988–1 (Flexicover)

*Printed and bound in Great Britain by
Redwood Burn Limited, Trowbridge, Wiltshire*

This book is dedicated
to the memory of my mother
Mrs Margaret Eileen Williams

Materials Engineering Practice

FOREWORD

The title of this new series of books "Materials Engineering Practice" is well chosen since it brings to our attention that in an era where science, technology and engineering condition our material standards of living, the effectiveness of practical skills in translating concepts and designs from the imagination or drawing board to commercial reality, is the ultimate test by which an industrial economy succeeds.

The economic wealth of this country is based principally upon the transformation and manipulation of *materials* through *engineering practice*. Every material, metals and their alloys and the vast range of ceramics and polymers has characteristics which requires specialist knowledge to get the best out of them in practice, and this series is intended to offer a distillation of the best practices based on increasing understanding of the subtleties of material properties and behaviour and on improving experience internationally. Thus the series covers or will cover such diverse areas of practical interest as surface treatments, joining methods, process practices, inspection techniques and many other features concerned with materials engineering.

It is to be hoped that the reader will use this book as the base on which to develop his own excellence and perhaps his own practices as a result of his experience and that these personal developments will find their way into later editions for future readers. In past years it may well have been true that if a man made a better mousetrap the world would beat a path to his door. Today however to make a better mousetrap requires more direct communication between those who know how to make the better mousetrap and those who wish to know. Hopefully this series will make its contribution towards improving these exchanges.

MONTY FINNISTON

Preface

The microprocessor revolution has now spanned over a decade in time; during that period the devices used have developed from rudimentary computing elements to complex, multifunction sub-systems on a single integrated circuit. The explosive growth of microelectronics has been both dramatic and traumatic. Many devices which emerged during this period were rejected by the market place, while others gained wide acceptance and have become standards for industry. We are now at a stage where fifth generation microcomputing systems are nearing fruition and promise computer architectures very different from the classic scheme envisaged by John Von Neumann.

As with all other rapidly developing technologies, industry has been slow to adopt microelectronics on a wide scale. The reasons for this are many and varied; industry tends to show a natural conservatism to change, particularly when existing equipment does not need replacing and functions perfectly well using older technologies. The rapid development by a large number of manufacturers of sets of devices which perform similar functions leads to the problem of which set to adopt. Engineers and technicians within industry have to learn a system which requires considerable investment of both time and money. All these factors contribute to a time lag of two to three years between a product being developed and it finding some acceptance.

In direct contradiction to this conservative approach by the established industries, new companies have been set up with the specific intention of exploiting the new technology. The growth in the home computing market is a prime example of a market developed as an outlet for microelectronics. The device manufacturers are extending into the construction of complete computing systems as a means of increasing the sales of their products.

The structure of a computing system presents unique problems when it fails to operate correctly and requires testing. Unfortunately the development of suitable test equipment has seriously lagged behind the growth of microelectronic devices. Early versions of test equipment were based on conventional instruments, modified to accept and present information from microprocessor based systems.

The information displayed was difficult to interpret and the instruments were difficult to set up. Improvements in design and in the ergonomics of the test equipment have evolved, to make them easier to use and to understand their results.

The techniques used to fault-find on microcomputer systems have themselves developed to the point where there is now an insistence from the product user to have the systems fitted with test facilities at source and not be left as an afterthought for the user to retrofit. Many of the methods used to test microprocessor systems require only simple additions which are considerably easier to fit when the system is being manufactured, but this may prove difficult after it has been constructed.

To test and repair a faulty system intelligently, the tester has to know the system being tested; knowledge is needed of what equipment to apply in a given situation; and a well-thought-out plan of attack should be followed. The first requirement is beyond the scope of this or any other book on test techniques, but the second and third requirements form the basis of this text. A well-formulated approach to testing any system can be established by developing a "troubleshooting tree" which attempts to reduce the testing burden to a sequence of tests which build on each other. This approach as a suitable vehicle for testing digital systems, along with the types of test equipment that should be applied for specific tests, form the main body of this book.

The book is intended for those who wish to discover the techniques and equipment which can be applied to the testing of microprocessor based systems. The material covered should find acceptance from practising test technicians and engineers who need to expand their knowledge in this rapidly growing field. Where possible, the book has been written as separate chapters on particular topic areas which can be read in isolation. The intimate relationships between certain sections of the material, however, do involve cross-referencing between chapters.

The bulk of the material has been taught for several years to full- and part-time students pursuing Higher Certificate and Higher National Diploma courses and to postgraduate students on short refresher courses.

The introductory chapter on microprocessor systems has been included for completeness and is not intended as a complete work on the subject. The chapter concentrates on certain topics, such as tri-state gates and address decoding which are fundamental to the operation of computer systems but which are often scantily covered in many of the available texts. The design of a microcontroller, suitable as the control element for many industrial processes, is outlined and the chapter concludes with a short discourse on programming levels.

The second chapter deals in a general way with some of the problems specific to bus structured systems and discusses many of the basic concepts used to test and isolate faults on computers. The view of a system as a kernel, surrounded by peripheral devices, is expounded, where the kernel has to be operative before any major tests can be applied. Chapter 3 introduces many of the basic principles that are applied when testing systems, particularly at first line maintenance level. The topics covered include stress testing and the "troubleshooting tree" as a formulated approach to fault finding.

The fourth chapter deals with the use of conventional test equipment such as oscilloscopes and digital voltmeters, as applied to testing microcomputers. The intention is to highlight their limitations and to introduce the need for test equipment specifically designed to analyse faults in microcomputers.

Chapters 5 to 8 deal with the instrumentation used specifically on digital equipment. Chapter 5 introduces the simplest test instruments and explains their uses. The chapter covers logic probes, logic pulsers, current tracers and logic comparators. Each of these hand-held tools is explained in terms of their operations and usage in typical test situations. The limitations of hand-held tools serves as an introduction to the following chapters on the more complex digital testing instruments.

Chapter 6 discusses logic analysers from their development as oscilloscope based instruments to the sophisticated equipment currently available. The types of display available are discussed along with their uses in test situations. Several of the ideas expounded in this chapter stem from projects undertaken with final year Higher National Diploma students and I am particularly indebted to Mr P. Davies and Mr S. Hunter for their contributions.

Chapter 7 covers signature analysis and its development from the transition counting technique. The mathematical basis for the method and its high probability of detecting errors are explained. The technique requires the system under test to be configured to accept it; this involves extra hardware in the system, along with built-in test programs. The majority of systems in use can however be retrofitted with these facilities, which are described in the chapter.

Chapter 8 deals with emulation as a test technique. It opens with a discussion of simulation as a means of mimicking system behaviour and expands into methods of controlling one system from another so that various levels of emulation can be applied. The term "emulation" is now applied to a wide variety of instruments from large development systems to self-contained test instruments; the rudiments of development systems are covered along with an outline of the types of free-standing emulators in use. This area of test equipment covers many of the previously outlined methods and is subject to wide disparity between instrument capabilities.

The penultimate chapter deals with the type of test programs used to check out parts of a system. These range from self-test programs which are executed whenever a system is switched on to programs which are only executed after a fault has appeared. Examples of test programs are given which may be used to check out the major components of a computer system.

The final chapter covers the testing of devices and systems peripheral to the computer system. Functional testing of a system is covered with an example of a data acquisition scheme configured with functional testing in mind. The use of go/no go indicators as a means of simplifying first line maintenance is discussed and applied to the data acquisition scheme. The chapter also covers the serial communication standard RS232C and the IEEE-488 parallel interface bus.

The main intent of the book is to bring together in one volume the major test methods in current use and as far as possible to explain them and describe their development from basic principles. Test equipment, in common with the systems they are applied to, evolve as better techniques are developed and as the systems they test become more complex. The book can under these circumstances only represent one viewpoint in time and the material contained within it will, in the course of time, be replaced by improved systems and techniques. The book does, however, provide a good starting point from which to enter this ever-increasing complex world and should provide the reader with a basis on which to judge future developments.

I wish to extend my thanks to Mr P. A. I. Davies and to Mr K. R. Webber of Gwent College of Higher Education, Newport, Gwent, for proof reading the manuscript and making helpful suggestions about the form and content of the material contained within this book.

CHART OF SYMBOLS

T transistor

D diode

R resistor

C capacitor

V_{cc} the most positive potential in a bipolar circuit (typically +5 volts)

V_{dd} the most positive potential in a CMOS circuit (typically +5 volts)

V_{ss} the most negative potential in a CMOS circuit (typically at ground potential)

V_{IL} input potential of a logic gate when in the logic "0" state

V_{IH} input potential of a logic gate when in the logic "1" state

V_{OL} output potential from a logic gate when in the logic "0" state

V_{OH} output potential from a logic gate when in the logic "1" state

V_N noise potential

N the number of sequences

P_x the probability of x

Contents

Chapter 1

Introduction to Microprocessor Based Systems

There are few, if any, electronic components that can match the prolific growth and diversity of use of the microprocessor. The first readily available device was released by the Intel Corporation of California in 1971 and was quickly followed by a multiplicity of similar devices from other manufacturing sources. The microprocessor attempted to implement, in a single integrated circuit, all the functions found in the *Central Processing Unit* (CPU) of a digital computer. The architecture of these first generation microprocessors was relatively rudimentary, when compared with a typical minicomputer, which was the smallest and cheapest digital computer then available. The original devices were followed by second and third generation components, which were significantly more powerful and complex than their forerunners. The increase in packing density and speed of operation can be directly attributed to continual improvements in the fabrication technology used to manufacture integrated circuits.

The availability of complex functional units as single integrated circuits has shifted the emphasis of design from circuits to systems. The components of the system have all the attributes of a digital computer with the functional detail of the system mapped onto them by a set of programs. The designer configures what is essentially a general purpose set of components to carry out the specific functions of a system by arranging for the computer to execute a list of instructions that implement the required system actions. The behaviour of the system is now controlled by the system's software, i.e. its set of programs.

A computer system is symbiotic in the sense that its electrical components (the hardware) and its programs (the software) are totally dependent upon one another and either is of little use without

1

the other. Particularly during the development phase of a microprocessor based product, this interaction often makes the problem of deciding whether a fault lies in the hardware or the software a difficult decision. To assist in clarifying the problem area, special purpose test equipment has been developed which ranges from simple instruments up to complex, multifunction test systems.

The skills demanded of engineers and technicians who have to design, maintain and repair these systems are necessarily wider than their traditional roles previously catered for. The education system has had to broaden its perspectives to cope with these needs, at a time when many of the fundamental principles underlying the design and testing of computer systems are themselves undergoing change. The widespread use of computing systems has brought into question the design of system software at all levels of programming. The manner in which many processors are connected together to form a distributed system and the overall control of such schemes remains largely unsolved. The testing of complex systems through built-in testing programs and the use of special test equipment has seriously lagged behind their application.

1.1 THE DIGITAL COMPUTER

Any digital computer contains the three essential elements shown in Figure 1.

The central processing unit (CPU) exercises control over the computer and internally carries out all arithmetic and logic operations. In a microcomputer, a microprocessor carries out the functions of a CPU.

The memory stores the lists of instructions (the programs) that are operated upon by the CPU. Each instruction is brought into the CPU and decoded by it to carry out the operations encoded into the

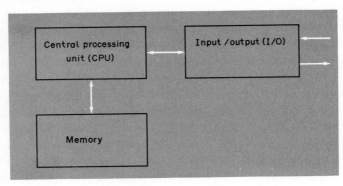

FIGURE 1   THE DIGITAL COMPUTER

instruction. Usually, a program will consist of a set of instructions, whose sequential execution by the CPU defines the system's operations. The memory is also used to store data values, such as the intermediate results of calculations, that cannot be conveniently stored within the CPU itself.

The *Input/Output* (I/O) stage allows the computer to communicate with the outside world. Under control of the CPU, data may be read in through an input port or sent out to some external device through an output port.

The type of external device connected to a computer will be determined by the application of the system. A general purpose computer will use peripheral devices, such as keyboards, *Visual Display Units* (VDUs) and printers, which allow human interaction with the machine. Digital computers used for control and instrumentation applications will have transducers connected to them to provide electrical analogues of process variables such as pressure, temperature and flow rate. Their output ports will connect to actuators such as stepper motors and control values to regulate process parameters. Human interaction with these systems is often only allowed through adjustment of a set point for a process.

The digital computer provides a general purpose structure, which, through the connection of suitable external devices and programming, can be used to fulfill a wide variety of tasks.

1.2 BUS STRUCTURED SYSTEMS

Information is conveyed from one point in a computer system to another over lines which carry only binary valued signals. Binary valued signals can only have two possible states or voltage levels, referred to as logic "0" and logic "1" states. A positive logic convention ascribes zero volts to the logic "0" state and typically +5 volts to the logic "1" state.

If information is transferred from one place to another over a single line by the sending end placing logic levels on the line as a sequential time series, then the form of communication is called serial transmission.

Figure 2 illustrates the transmission of the 8-bit binary word 01001100 over a serial transmission line. The word is shown being sent with the *Most Significant Bit* (MSB) transmitted first; in practice, it is usual to send a word of data in a reverse order, with the *Least Significant Bit* (LSB) leading. Serial transmission uses the least number of communication lines possible, but takes N clock periods to send a word of N binary digits or bits.

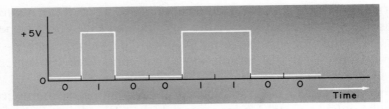

FIGURE 2   SERIAL TRANSMISSION

An alternative form of transmission uses a separate line to convey each of the N bits, with the lines arranged in ascending binary weighting. This form of transmission is called parallel transmission and collectively the set of lines over which the information is sent is called a *bus*.

Information is sent over the bus in bit parallel, word serial form with each word taking only a single clock period to transmit from the sending end to the receiving end. Each piece of data is read from the bus in synchronism with the system's clock; Figure 3 depicts the word 01001100 being sent over an 8-bit bus at some instant in time t.

To achieve high rates of transmission, transfers within a digital computer use buses. The penalty paid for implementing such a bus structured system is an increase in the complexity of the computer's components, but this is more than compensated for by the increase in information throughput.

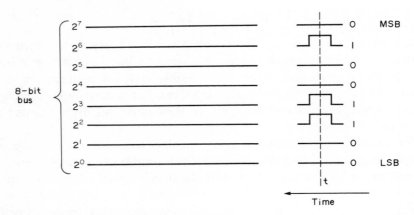

FIGURE 3   PARALLEL TRANSMISSION OVER A BUS

A computer uses three busses to transfer one item of data from one point to another, as illustrated in Figure 4.

Data transactions are controlled by the CPU which initially specifies the location in the system which it wishes to read data from

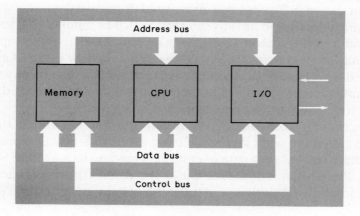

FIGURE 4   A BUS STRUCTURED DIGITAL COMPUTER

or write data to, by placing the address of the location on the *Address Bus*. In a small system, only the CPU will be capable of issuing addresses and for this reason the address bus is said to be unidirectional. The number of lines or width of the address bus defines the total number of unique locations that can be specified by the CPU. The majority of microprocessors have 16 lines in their address bus which enables the CPU to specify any one of 2^{16} (65,536) separate locations. It is common to refer to any multiple of 1024 (2^{10}) as K, thus giving the system a direct addressing capability of 64K.

Having placed an address on the address bus, the CPU now has to inform the other elements in the system about the type of operation being carried out. An operation may be a read or a write from or to a memory location or from an input port or to an output port. The CPU qualifies the operation by enabling lines in its *Control Bus* which selected either memory or input/output and specifies whether the operation is a read or a write. Not all lines in the control bus emanate from the CPU; when reading from a memory device, say, the device may not be able to repond in the time expected by the CPU, and it would under such circumstances send a control signal back to the CPU informing it that it has to wait until the data requested can be made available. Some signals in the control bus are sent by the CPU, while others are received by it. The term quasi-bidirectional has been coined to describe the collective bidirectional nature of the control bus, although any one line in the bus is unidirectional.

Having specified an address within the system and qualified the type of operation using the address and control busses, the CPU then either sends data to the specified location or receives data from the location over the *Data Bus*. Any of the elements of the computing system shown in Figure 4 can either send or receive data over the data bus which thus has to be fully bidirectional. The number of lines in, or the width of the data bus provides a means of classifying computers. The majority of microprocessors have 8 lines in their data bus and are termed 8-bit microprocessors; some microprocessors and most mini-computers employ 16 lines in their data busses and are called 16-bit microprocessors or 16-bit machines.

Collectively 8 bits are referred to as a byte and the width of the data bus is called the word size of the computer. An 8-bit microprocessor thus transfers bytes of data over its data bus and its word size is also a byte. The word size of the computer also defines the size of each of its memory storage locations and the width of its input and output ports.

1.3 MEMORY MAPPED SYSTEMS

The general, bus structured computer shown in Figure 4 indicates an input/output stage completely segregated from the memory of the system. The majority of minicomputers and many microcomputers treat input and output ports as if they were simply memory locations. As far as the CPU is concerned, reading from an input port is treated in precisely the same manner as if the CPU were reading the contents of a memory location. Similarly, writing to an output port is treated by the CPU as if it were writing to a memory location. The advantage in using memory mapped I/O is a saving in the number of control signals needed in the control bus, because a separate control signal indicating a reference to an I/O operation is not required. Since all data transfers are treated as memory references, a control signal indicating a memory reference can also be dispensed with. The disadvantage of memory mapped I/O is that the input and output ports occupy addresses that cannot now be used for memory devices and the total memory space has to be shared between memory and I/O.

The alternative to memory mapped I/O is the awkwardly named I/O mapped I/O scheme used in the microprocessors manufactured by Intel and Zilog, where separate control signals are used to access memory locations and I/O ports. Besides adding to the complexity of the control bus, microprocessors employing this technique use sepa-rate instructions in their instruction sets to handle I/O transactions, which are generally inferior to the access and manipulation instruc-tions that apply to memory locations. The advantage in using an I/O mapped I/O scheme is that the address spaces of a system's memory and I/O ports are completely separate, thus allowing the total memory space to be used by memory devices.

The architectures of microprocessors that employ these different I/O mapping techniques also tend to differ. In general, micro-processors that make use of I/O mapped I/O also tend to have a relatively large number of internal storage registers for temporary data storage, while devices that use memory mapping for I/O stages have few internal storage registers, which necessitates storage of virtually all data in memory external to the CPU itself.

1.4 TRI-STATE DEVICES

Each of the functional blocks represented in the bus structured computer system can place data onto the data bus upon request from the CPU. Each block connects onto the data bus through logic gates, which if of conventional construction could only output a logic "0" or a logic "1" level onto the data bus line to which it was connected. Every line in the data bus would have many such gate outputs connected onto it, some of which could be at one logic level while the rest were in the opposing logic state. Under these conditions the actual state of the line would be uncertain and a bus conflict would be said to exist where one output device was attempting to pull the line to one level while another tried to pull it to a different potential.

The basic logic gate structure which provides an active source of current at its output and an active device to sink current from its output terminal is the 7400 TTL NAND gate.

The totem pole output stage of the 7400 gate always has one of its transistors turned on and the other off. If it were used as a data bus

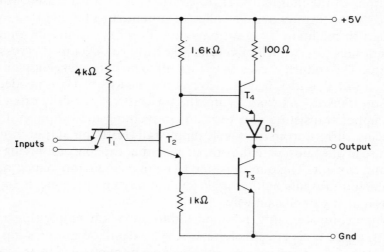

FIGURE 5   7400 TTL NAND GATE

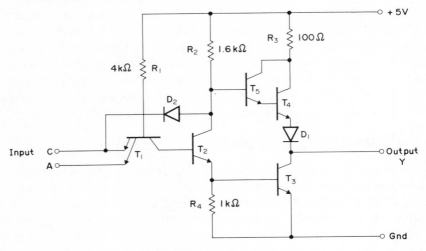

FIGURE 6 THE TRI-STATE GATE

line driver, it would cause bus conflicts; the gate can, however, be modified by the addition of two extra components to provide a gate that has three possible output states.

To the 7400 gate circuit has been added diode D_2 and the upper transistor of the totem pole output stage has been converted to a Darlington pair by the addition of transistor T_5. When the control input C is connected to ground (logic "0"), diode D_2 clamps the collector potential of transistor T_2 at the forward voltage drop across D_2 above ground, approximately 0.6 volt. One of the base-emitter junctions of the multi-emitter transistor T_1 is forward biased, which causes T_1 to turn on and its collector potential to fall below a level needed to maintain T_2 in an on state. T_2 turns off and its emitter current falls to zero, which reduces the voltage developed across R_4 to zero. This in turn causes T_3 to turn off. The potential clamp on the collector of T_2 is insufficient to overcome the forward bias needed to turn on transistors T_5 and T_4 and the forward voltage drop cross D_1. The upper transistors also turn off, preventing the output Y from sourcing current from the 5-volt supply rail. Both the output transistors are in off states, with the output Y neither capable of sourcing nor sinking current. The output is said to be in its high impedance state, in which it presents neither logic level to any line it connects to nor loads it in any significant way.

When control input C is connected to the 5-volt supply (logic "1" state), diode D_2 is effectively reverse biased and does not clamp the collector of T_2. Transistor T_1 can now only be controlled by the input A through grounding it to turn T_1 on or by taking it to the 5-volt rail to maintain T_1 in an off state. The gate now behaves as a logic inverter, with the output Y always in the opposite logic state to the input A.

The gate shown in Figure 6 has three possible output states. With the control input C at a logic "1" level, the output Y will be in one of two possible binary states depending upon the state of input A. When C is taken to a logic "0" level, the output will be in a high impedance state, which isolates the gate from the line to which it is connected. Gate structures of this type are called tri-state gates and are used extensively in computer systems to connect devices onto the bidirectional data bus lines.

Tri-state devices are fundamental to the manner in which the CPU exercises control over data transfers on the data bus. On any device that can output data onto the data bus there will be found one or more inputs labelled as $\overline{CE}$ (not chip enable), $\overline{CS}$ (not chip select), or $\overline{OE}$ (not output enable). These control inputs function in a corresponding way to the control input C of the gate shown in Figure 6 and enable or disable all the data bus drivers in the device. Thus if the device were a byte wide memory chip, then its control input would either place all 8 output lines into a high impedance state or allow them to place binary levels onto the lines to which they were attached. Conventionally, control signals in a microprocessor system use a negative logic standard, indicated by a bar written over the signal's mnemonic. The true or active state of such signals is when they are at ground potential, which means for the memory device that it can place data onto the data bus when its $\overline{CE}$ input is taken to a logic "0" state. In its quiescent condition, the CPU places all devices connected onto the data bus in their high impedance state so that no device can place data onto the bus. When the CPU wishes to read a memory location or an input port, it selectively enables only that device, keeping all others disabled. The selected device places data onto the bus, from which it is read by the CPU. The CPU then disables the selected device to return the bus to its quiescent state. Information flow within the computer is controlled through the CPU selectively enabling one device at a time to place data onto the data bus, including its own data bus line drivers.

1.5 ADDRESS DECODING

A typical computer system will contain many memory devices to store programs and to act as a temporary data storage area. It will also have a number of input and output ports to allow communication with external devices. Every memory device, input and output port has to

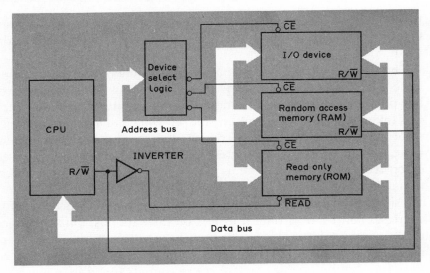

FIGURE 7   A SIMPLIFIED ADDRESS DECODING SCHEME

be selectively enabled by the CPU when it is required to take data off the data bus or place data onto it; clearly the CPU cannot directly provide all the separate select signals for every possible device that could be connected into a system. The select signals have to be decoded from address bus information and qualified by control bus signals.

Generally the components used in microcomputers do not contain a single storage location or an individual input or output port. Memory devices are organised as blocks of 1024 (1K) locations and contain internal decoding logic to permit access to an individual location. A typical memory chip used in a microcomputer will contain 1, 2 or 4K byte wide storage locations; the individual selection of a location is determined by decoding logic which decodes the specific address from the states of the address bus lines. A 1K device will have the lower 10 address lines directly connected to it (A_0 to A_9) to achieve the required decoding for a specific location.

There are exception to the general case where memory devices may contain less than 1K of total storage, or be organised with less than 8 bits per storage location. The general philosophy of device selection to be outlined remains the same, and only the interconnection details change to suit the type of devices in use. In small or low cost systems, individual input and output ports may be used which require their addresses to be uniquely decoded from the address bus. To simplify the address decoding logic, the I/O ports are usually located in a block of memory addresses, any of which will enable the port in question.

The trend in microcomputers is, however, to use programmable I/O devices which contain several locations associated with I/O functions, and to use internal decoding logic to select an individual location.

The higher order address lines are used to generate the chip enable signals for the various memory and I/O devices in a system. Figure 7 shows the address bus split between memory and I/O devices and device selection logic. The lower order address lines are directly connected to each of the memory and I/O components and the upper order address lines connect to the selection logic. To access an individual location in the system, the total address bus bit pattern is used; the lower order lines specify a location within a block of addresses and the upper order lines specify the block of addresses. Using this technique, a device may be located anywhere in the total memory space, its precise location being determined by the connection between its chip enable input and an output from the device selection logic.

There are two types of semiconductor memory in general use: *Random Access* or *Read/Write Memory* (RAM) can be written to or read from, while *Read Only Memory* (ROM) can only be read from by a microprocessor. ROM is non-volatile, which means that its contents are not lost when power is removed from it, while RAM is volatile. ROM is used to store programs where it would be inconvenient to have to reload the program every time the machine was switched off. The majority of instrumentation and control applications fall into this category. To prevent bus conflicts, the enabling of ROM will often be qualified by a control bus signal which indicates a read operation. In the simplified scheme shown in Figure 7, control bus signals are drawn in heavier lines for clarity. A typical control bus signal which indicates whether the operation being carried out is a read or a write is shown as $R/\overline{W}$. This mnemonic is used to signify that a read operation is specified when the $R/\overline{W}$ line is at a logic "1" level, while a write operation corresponds to a logic "0" on this line. The majority of device enabling signals are active low, i.e. the signal is true when it is at a logic "0" level, which requires that the $R/\overline{W}$ signal be inverted before being applied to the ROM. A common practice used to indicate that signals are active low is to draw the line on the device with a circle as shown in Figure 7 for the $\overline{READ}$ control input to the ROM.

RAM devices have the $R/\overline{W}$ control signal, which is used internally to determine the type of operation requested, directly connected to them. Thus if the $R/\overline{W}$ line is at a logic "0" level when the device is enabled through its chip enable input being taken low, then a write operation is being carried out and data on the data bus will be loaded

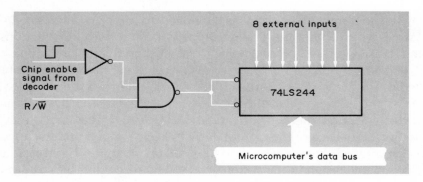

FIGURE 8   SELECTING A SIMPLE INPUT PORT

into the location specified by the lower order address lines connected to the RAM. If the R/W̅ line is at a logic "1" level when the RAM is enabled, a read operation is inferred and the contents of the location decoded off the lower order address bus lines is placed onto the data bus.

Programmable I/O devices are treated in a memory mapped system in the same manner as RAM with an input port being read when the device is selected and the R/W̅ control line is at a logic "1" level and an output port loaded when the R/W̅ line is at a logic "0" level. Where simple *Small Scale Integration* (SSI) or *Medium Scale Integration* (MSI) logic devices are used as input and output ports, the enabling signals from the device selection logic have to be qualified using external logic with the R/W̅ signal.

A device often used in small systems as an input port is the 74LS244 quad bus transceiver. The device may be reconfigured as shown in Figure 8 to act as an 8-bit input port, through which the states of 8 external signals may be read into the microcomputer. The 74LS244 has tri-state outputs, so that when its enabling inputs are at a logic "1" level its outputs are disabled and do not load the data bus. An active low chip enable signal taken from the system's device selection logic is used to enable the 74LS244 and allow it to transfer its input states onto the data bus lines. If the programmer had inadvertently written an instruction which attempted a write operation to this input port, unless the chip enable was qualified with the R/W̅ signal, a bus conflict would arise. The microprocessor would attempt to place data onto the data bus and the input port would be enabled which would also attempt to place data onto the bus. This situation can be avoided by the logic illustrated in Figure 8 where the chip enable signal is gated with the R/W̅ line so that the 74LS244 can only be enabled for a read operation. An inadvertent write operation would prevent the enabling of the input port and thus prevent a bus conflict situation.

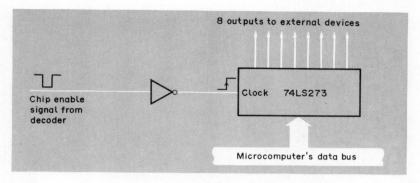

FIGURE 9   A SIMPLE OUTPUT PORT

Notice in Figure 8 that the only address information used is the upper address lines to the device selection logic. This simple input port would thus appear in a memory map of the system at every address covered by the block select signal, and no unique address will exist for it.

Latching data off the data bus to a simple output port is shown in Figure 9.

The 74LS273 contains 8 D-type latches; the Q outputs connect to devices external to the microcomputer, while its D inputs connect directly onto the microcomputer's data bus. D-type latches use a positive going edge to clock information on their D inputs through to their Q outputs. The active low chip enable signal is hence inverted by a logic gate to clock data bus information to the device's outputs. Notice the absence of qualifying logic to uniquely define a "write" operation; in the case of an output port, a bus conflict cannot arise, because an inadvertent read from such a port would only cause the CPU to read the inactive or quiescent state of the data bus. Qualifying logic may, however, be included to prevent the outputs of the 74LS273 from changing state due to an erroneous read operation being specified and a typical scheme is shown in Figure 10.

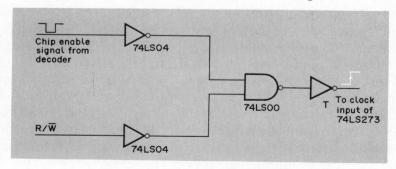

FIGURE 10   SIMPLE OUTPUT PORT QUALIFYING LOGIC

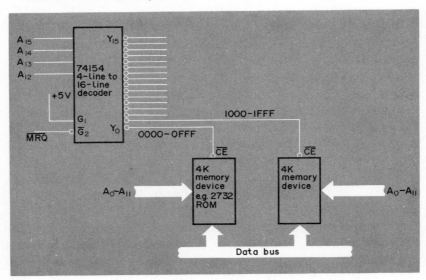

FIGURE 11   ADDRESS DECODING ON 4K BOUNDARIES

In the majority of microcomputers a single MSI logic decoder is used for the device selection logic. For an 8-bit microprocessor, with 16 lines in its address bus, the total of 64K of memory space may be partitioned into 16 4K blocks using a 74154, 4-line to 16-line TTL decoder device. Figure 11 illustrates such a device being used with the upper four address bus lines (A_{12} to A_{15}) decoded into 16 outputs on 4K address boundaries.

The bit pattern on the 4 input lines causes one of the 16 output lines to change from a quiescent logic "1" state to an active low, logic "0" state, provided that the two enable inputs G_1 and $\overline{G_2}$ are in their active states. Thus if the device were enabled and the bit pattern 0000 were applied at its inputs, the $\overline{Y_0}$ output would be in the logic "0" state, while all the other outputs would remain in their inactive logic "1" states.

In Figure 11 a memory request control bus signal $\overline{MRQ}$ is shown controlling the 74154 decoder. This diagram illustrates an I/O mapped I/O system because a memory request signal has been used to qualify the selection logic; a similar decoding scheme would apply to the selection of I/O ports with the memory request signal replaced with an I/O request signal. In a memory mapped I/O system, both memory and I/O devices would be enabled through a common address decoder, which would be either permanently enabled by connecting its enable inputs to the supply rails, or enabled by some synchronising signal produced by the microprocessor. The scheme shown in Fig. 11

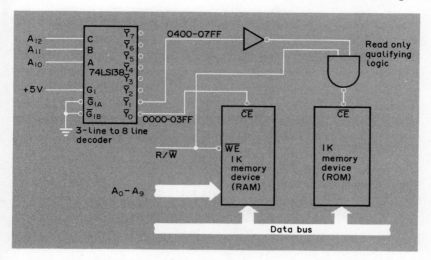

FIGURE 12   8K MEMORY SPACE DECODED ON 1K ADDRESS BOUNDARIES

could be modified for a memory mapped system by tying G_1 directly to the +5 volt supply rail and $\overline{G}_2$ directly to ground.

Many small systems, such as dedicated industrial controllers, do not require the total memory space available. The block select signals may be required on boundaries less than the 4K memory spaces illustrated in Figure 11. If a system's requirements can be fulfilled in less than 16K of memory and 1K memory devices are to be used, the scheme shown in Figure 11 may be altered by using the address lines A_{10} to A_{13} as inputs to the 74154 decoder. Sixteen outputs are then available from the decoder on 1K boundaries and the lower ten address lines (A_0 to A_9) are directly connected to each of the 1K memory devices.

An alternative decoder, found in many small systems, is the 74LS138, 3-line to 8-line decoder. Figure 12 illustrates an 8K memory space decoded on 1K address boundaries, where the address lines A_{10}, A_{11} and A_{12} are used to provide eight 1K address block enable signals. Notice in the scheme that the chip enable signal to the ROM device has been qualified with the "read/write" control signal to prevent bus conflicts occurring.

By a suitable choice of address decoding device and address lines, the memory space may be partitioned into any convenient size. Where the memory chips in use are organised to contain less than the word size of the computer, they may be paralleled to form a memory block with a width equal to the required word size.

The 2114 RAM device shown in Figure 13 is internally organised as 1024, 4-bit storage locations, which entails using two such devices in

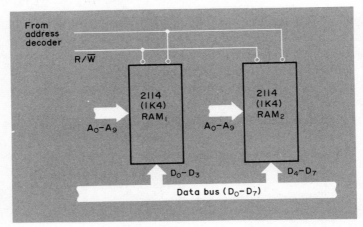

FIGURE 13   PARALLELING MEMORY DEVICES

parallel to provide byte wide storage. The lower order address lines and control signals connect to each device in parallel, while the data bus is split between them. The 4 lower data bus lines (D_0 to D_3) connect to one RAM, while the upper 4 lines (D_4 to D_7) connect to the second RAM device. When a byte of data is written into the RAM, the lower nibble (4 bits) is written into RAM_1 and the upper nibble into RAM_2. Similarly, a read from RAM would cause the upper nibble to be read from RAM_2 and the lower nibble from RAM_1. The 8-bit microprocessor is sent a byte of data in total and views the RAM as if it were a byte-wide device. Similar reasoning applies where components are organised as bit-wide storage devices, the only difference being that 8 such devices are paralleled to form a byte-wide memory storage block.

1.6 A MICROCONTROLLER

A microcontroller is a small microcomputer system that can be applied to control and measurement tasks. Its program is dedicated to carry out a limited number of specific functions such as reading transducer signals, manipulating data values, and setting output control signals. Non-volatile memory is needed which will involve the program being stored in ROM, with a small amount of RAM included to hold temporary data values. Microcontrollers differ from general purpose computers largely in their relative use of ROM and RAM memory. A general purpose machine would use a large amount of RAM, because the programs it operates on may be replaced at frequent intervals, which requires the use of read/write memory, and only a proportionately small amount of ROM for the storage of often

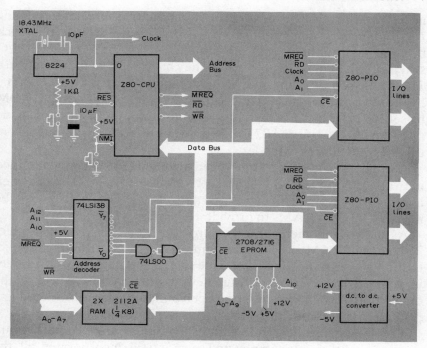

FIGURE 14   A Z80 BASED MICROCONTROLLER

used, fixed programs. A microcontroller would contain a relatively
large amount of ROM to store its fixed programs, and only a small
amount of RAM to act as storage for temporary data values and as a
stack area. A microcontroller would also require a larger number of
I/O lines through which to communicate with external devices while
the general purpose machine would only provide a limited number of
special purpose I/O lines, such as serial data links, to connect to
VDUs and printers.

Microcontrollers tend to be almost minimal configuration systems
and can generally be contained on a single printed circuit board.
Expansion is catered for by the inherent bus structured nature of the
scheme, with extra memory or I/O ports constructed on other circuit
boards, which communicate with the microprocessor through edge
connectors and backplane wiring. Figure 14 illustrates a microcontroller based on the Zilog Z80 microprocessor, which could be
used as the basis for a wide variety of industrial controllers.

Associated with the microprocessor is a system clock, constructed
using an Intel 8224 clock chip. This provides a 2 MHz clock frequency
and all operations carried out by the microprocessor are tied to this.
In place of simple I/O ports, programmable devices have been

included to increase the flexibility of the system. These devices, Zilog *Programmable Input/Output components* (PIOs), require the clock be connected to them. PIOs are usually used in an I/O mapped I/O scheme, but in this microcontroller they have been memory mapped. This saves incorporating an extra address decoder for the I/O ports.

Facility has been included to use either a 1K ROM or a 2K ROM by setting two links. As the system stands, the EPROM can be made a 4K device, such as the 2732 EPROM, by altering the outputs from the 74LS138 address decoder.

A small amount of RAM has been included (256 bytes), contained in two 2112A static RAM devices. RAM provides a stack area as well as temporary storage for calculated data values. The devices could be replaced by 2114 RAMs to provide 1K of read/write memory.

As shown, the microcontroller provides 1 or 2K of ROM storage, 256 bytes of RAM and 32 uncommitted I/O lines. Including the d.c. to d.c. converter needed to provide the 12 volt and −5 volt supplies used by a 2708 EPROM, only ten devices are needed to construct the microcontroller and several have been built on a single Eurocard size printed circuit board.

The microcontroller demonstrates that not all microcomputers are necessarily large or complicated, yet it still incorporates all the essential elements found in digital computing systems of: system clock; CPU; memory; and I/O stages. The simple address decoding scheme caters for up to eight 1K blocks which may be occupied by ROM, RAM, or I/O. On power up or when reset, the Z80 forces its program counter to zero, so that in a dedicated scheme EPROM would be placed at the lower end of the total address space available.

1.7 PROGRAMMING LEVELS

A computing system may generally be programmed at several levels. To understand why several levels of programming languages have developed and been found desirable, it is necessary to examine the operations carried out by a CPU during the execution of a typical instruction. Each instruction is stored within the computer in a specific form, which typically would be OPCODE, OPERAND, OPERAND. The OPCODE or operation code part of the instruction informs the CPU of the type of operation required, such as ADD, SUBTRACT. MOVE and may also contain information about the locations of the numbers to be operated upon or the OPERANDS. In the typical format given, the numbers to be operated upon are stored as part of the total instruction. Several instruction formats are found necessary to cover the complete range of instructions that the computer can execute. An alternative to the format given may be OPCODE, ADDRESS where the instruction causes the program to jump to a non-contiguous memory location, specified by ADDRESS, to continue its operation. Each part of an instruction has to be stored within a machine as a series of binary numbers; the machine differentiates between command numbers (opcodes) and true numbers (operands or addresses) by having those numbers placed in specific sequences within its memory that conform to its instruction set formats.

In the majority of computers, this is the lowest level at which an operator can program the machine. It is called MACHINE CODE programming and consists of loading into memory the relevant binary numbers that the machine accepts as opcodes and the data values that are to be associated with each instruction. To ease the programming task, facilities usually exist to input the numbers in a higher order number system that bears a direct relationship to binary, such as *Octal* or *Hexadecimal*. As an example of programming at this level, say we wish to add the decimal number 12 to a register within the CPU, which we shall call the A register. The machine may accept the code C6 in hexadecimal as the opcode for the operation. The instruction will have to be loaded into two consecutive memory locations as C6 followed by OC, where OC is the hexadecimal equivalent of the decimal number 12. When the instruction is executed, 12 will be added to the previous contents of the A register to form a new result.

While this machine code level is usually the lowest available to a programmer, computers actually operate on a lower level of programming again, called MICROCODE. Every instruction in the machine's repertoire, such as the ADD instruction previously cited, has to be broken down into a series of sub-operations which route data from one point in the system to another and cause the desired actions to take place. In the ADD instruction example, the operand has to be brought into the CPU from memory and placed in some temporary storage location; the *Arithmetic and Logic Unit* (ALU) has to be set up to carry out the ADD operation; the two operands have to be routed into the ALU and then the result routed back to some specified register. Every instruction has to be broken down into its corresponding sequence of sub-operations through the microcode stored within the CPU. This fundamental level of programming defines the machine's instruction set at its machine code level, and is seen as difficult and time consuming.

Certain devices are manufactured which can be microcoded by a user, such as bit-slice microprocessors. The primary reason why a user would want to perform this arduous task occurs where he wishes the device to emulate an existing machine, by microcoding the device to carry out the same instruction set as an existing machine, or where the instruction set required cannot be obtained from any existing machine. External to the CPU, microcode is transparent to a user, who can only perceive the macro-level machine code operations being carried out.

Generally, unless a user specifically wishes to implement his own instruction set, the lowest level of programming he will have recourse to will be machine code. Machine code programming is both tedious and error prone, because each instruction has to be entered into the machine as a series of numbers, only one opcode of which has to be wrong to cause drastic consequences. If a data value is wrongly entered, then a so-called "soft error" will occur and the wrong result ensue. The machine will not go seriously wrong, but the results may be unexpected.

To alleviate the difficulties associated with machine code programming, ASSEMBLER programs were developed which allow the user to program his machine using MNEMONIC equivalents to the machine code opcodes. In place of entering the numbers C6 and OC for the addition operation, he could now write ADD 12 as an instruction. The first benefit to the user is a reduction in hard errors where opcode values are wrongly entered. If, for example, the mnemonic UDD was incorrectly entered, then error checking routines within the assembler package would detect it and flag an error message to the operator. Mnemonic codes such as ADD, SUB, AND and MOV are commonly used to indicate the type of operation to be carried out.

Programs written in assembler form have a one-to-one correspondence with machine code instructions and are thus as efficient as them. Further user benefits that accrue from using an assembler are, the program may be written using labels for addresses and names for data values, rather than specific values. The program start address may be specified only during the assembly operation, so that a program may in one case be started at one address, but if that same program segment is later used in another program, its start address may be made some different value. The assembler program is then said to be RELOCATABLE because it can be located at any suitable address during the assembly process where it is turned into its machine code equivalent.

The instruction set of any computer is usually restrictive in the scale and type of operations that it can perform. Arithmetic instructions are restricted to integer values of a fixed size range, which for a typical

8-bit microprocessor will be numbers in the range -127 to 127. Where irrational numbers over a wide range have to be handled, programs have to be built up from the basic instructions in the instruction set to handle them. Trigonometric functions have to be evaluated using some series approximation for the function which has to be implemented using a sequence of machine code instructions. Where irrational numbers and complex mathematical functions have to be handled and evaluated, the machine will have a suite of programs stored in it that can be called up to provide the required facilities. Specific locations in memory will be reserved for the storage of the operand values for these functions which when called can expect to find the numbers stored at those locations in specific formats. Typically an operand will be stored as a MANTISSA and EXPONENT in three or four bytes of memory.

Even with these extra facilities added into a machine's program repertoire, programming at assembler level still remains a laborious and detailed task. To overcome the problems of programming at this level, higher level languages such as FORTRAN, BASIC and PASCAL were developed. The idea behind such languages is to make the task of programming a computer more human orientated than machine orientated. The so called HIGH-LEVEL languages attempt to achieve this by making instructions conform to English-like expressions and through the use of acceptable mathematical symbols such as / for the division operation. The majority of high-level languages are designed around a few programming constructs such as the IF . . . THEN . . . ELSE or WHILE . . . DO forms. For every such language there exists a set of syntax diagrams which define the form that instructions can take and which in most cases have to be rigidly adhered to. While spaces may be freely placed in instruction forms, other symbols used to define the syntax of an instruction such as commas or semicolons, or in some cases the alphabetic character used to start the name of a variable, have to be correctly placed or used, otherwise the program will not operate.

A computer can only operate on machine code instructions and so to make a program written in a high level language run on a machine it has to be translated into equivalent sequences of machine code instructions. This translation operation is not particularly efficient in terms of the size of the resulting machine code program, which almost always could have been written to execute more rapidly and using less instructions directly in machine code. This disadvantage is, however, more than offset by the much shorter time in which the high-level version of the program can be written compared to a machine code version.

The translation operation can be carried out in one of two basic ways using either a program called an INTERPRETER or one called a COMPILER. Interpreters and compilers are not single programs but suites of programs that interact to carry out the translation operations. The manner in which the two types of translation are carried out are fundamentally different in several important ways. An interpreter takes one line of a high-level language program and translates it into a machine code equivalent sequence of instructions which are then executed. Having completed one line, the interpreter then moves on to the next line in sequence and translates and runs that. No machine code program for the complete program ever exists within the computer, so that the next time the same program is run the interpreter has to translate it again line by line. Interpreters are particularly inefficient where a repetitive sequence of operations is carried out at several points within a program. Ideally, the sequence should be removed from the main body of a program and written once as a subroutine and called up when desired. The program is then most efficient in terms of the amount of memory space occupied. Interpreters are also relatively slow because each line of a program has to be separately interpreted every time it is encountered.

Alternatively, a compiler translates a complete program into a machine code equivalent, which is stored, and can be run in its machine code form. The compiler is hence used once to produce the machine code program and plays no further part when it is run. The compilation process is a relatively long exercise, taking several tens of minutes to complete for even simple, short programs and involves a suite of long, complex programs that cannot all reside in a computer's memory at the same time. Thus some form of external bulk storage such as floppy discs is needed to store the compiler and this is loaded into the computer memory in parts, each of which carries out specific operations The advantage of a compiled program over an interpreted program is that the compiled version exists as a complete machine code program and will hence run faster than the interpreted version. Certain well-designed compilers will search a program for common blocks and remove them as subroutines which can greatly reduce its final size.

A wide variety of high-level languages exist, most of which have been designed to satisfy the needs of specific interest groups. Those working in Artificial Intelligence will be more likely to use the LISP language or one of its offshoots, while someone interested in business applications will tend to use COBOL. The major differences between most high-level languages is the manner in which they represent, store and manipulate information fed into them. Any high-level language still only represents one level in a hierarchy of programming forms; there are envisaged levels of programming above those currently in use which would make the task of programming a computer relatively easy in comparison to today's high-level languages. However, using existing systems entails some translation process from any level down to machine code which is the only level (excluding the transparent microcode level) at which a digital computer can operate.

Chapter 2

System Testing Problems

Digital computing systems present some unique fault-finding problems due to their organisation and structure. In common with other electronic systems, the components used to construct a computer will fail through a variety of mechanisms and lead to partial or complete failure of the system. The architecture of the computer system, however, often makes isolation of the faulty component a difficult and time-consuming task.

A large proportion of the circuits used in a computer will be constructed using *Large Scale Integration* (LSI) fabrication techniques through which complete sub-systems of the computer will be supplied as single components. The microprocessor itself is an example of such an LSI component. As more and more circuitry is fabricated into a single device, the problems of functionally testing that device increases dramatically. Any LSI device only undergoes limited testing before being put into service, and faults may manifest themselves only after field usage, due to patterns occurring that were not used to test the device or component parameter changes that cause uncharacteristic device behaviour.

Satisfactory operation of a computer system requires that the software run on the system is fault-free as well as the hardware. This is often not the case, and programs that were thought to be completely functional may produce false results under certain circumstances. A program may perform correctly when fed certain test patterns but fail in use because an unforeseen pattern is produced which it cannot handle. Allied to this type of fault are those programs which through poor design do not tolerate an unexpected input from an inexperienced operator and which causes an apparent system failure.

Beyond these types of fault are those that develop within a system after it has been in use for some time. Generally, these will be electrical faults due to component failures or, less frequently, program faults that develop due to corruption of the stored instructions.

2.1 HARDWARE OR SOFTWARE?

The symbiotic nature of a computing system presents an immediate problem when a fault occurs in it. Unless the fault is an apparent one, the fault locator has to decide whether it lies in the hardware or is a bug in the software. This problem is often not an easy one to answer, because the nature of the fault may be such as to prevent simple diagnostic tests from being carried out. A fault on the micro-processor's control bus may prevent any program from running, while a fault in the operating system software of a microcomputer may also prevent any test programs from being loaded and run.

For small, microcontroller-type systems, the simple expedient of manually switching into circuit a set of test programs that run in the same memory space normally occupied by the main programs will often indicate the nature of the fault. If the test programs do not run, then a hardware fault is likely, while the complete test sequence will exercise the major functional parts of the system and thus indicate whether a fault lies in the main programs. Several approaches may be taken as to the manner in which such a test ROM is incorporated into the system. The simplest method is to unplug the system ROM and replace it with the test ROM in the same socket, which at least makes use of the system hardware that would normally be used, i.e. the interconnections between the CPU and the system ROM socket. Alternatively, a separate test ROM may be constructed as part of the system and a switch used to connect the chip enable signal from the system ROM to the test ROM when tests are to be carried out. These are by no means the only ways in which test facilities can be built into a system, and other variations on this theme will be discussed in later chapters.

This basic problem of determining if a fault lies in the hardware or software does highlight the fact that manufacturers of microprocessor based systems have seriously lagged behind in including test facilities within their machines, which could greatly simplify a tester's dilemma.

2.2 TIME SEQUENTIAL, BIT PARALLEL DATA

At any time, the complete state of a microcomputer system is contained on a large number of lines. In an 8-bit microcomputer, the address currently being accessed is represented by the state of 16 parallel lines, the *Address Bus*. Similarly, the data being taken to or retrieved from that address occurs as a bit pattern on 8 parallel lines, the *Data Bus*. To determine the type of operation being carried out requires that the state of several lines in the *Control Bus* is also known. Thus the state of many lines has to be known before all the information about one bus transaction is available. Any one piece of information may only exist on a bus for a very short period of time. A byte of data read from memory may only be placed on the data bus for a single system clock period, which may be as short as 250 nanoseconds.

Information is moved around the microcomputer in this time sequential, bit parallel form so that one piece of information may only exist within the system for a short space in time before being replaced with the next. Conventional test equipment, such as an *Oscilloscope*, is too restrictive to play any major role in detecting and displaying information in this form and recourse has to be taken to special instruments that have been developed specifically for the purpose of capturing and displaying the data from bus structured systems.

2.3   BUS MULTIPLEXING

The restrictions placed on the package size of LSI devices such as microprocessors has meant that insufficient pins are available on the package for all the signals produced by the device. The manufacturer then has to use some pins on the package for several functions which entails multiplexing signals onto them in a time sequence. The problem of insufficient pins is greatly exaggerated in 16-bit microprocessors such as Zilog's Z8000 or Intel's 8086. The same problem occurs in 8-bit microprocessors, and cases in point are Zilog's Z80 which has to use the control signals $\overline{M1}$ and $\overline{IORQ}$ together as an interrupt acknowledge signal and Intel's 8085 which time multiplexes the data bus with the lower 8 bits of the Address bus.

Bus multiplexing increases the complexity of testing because it has to be decided at any moment what information occupies the multiplexed line or lines. Again, conventional test equipment is totally unsuited to demultiplexing the information from such lines and special test equipment has to be used.

2.4   THE DEVICE TESTING PROBLEM

To bring the problems of testing LSI devices into perspective, the time taken to test a representative device will be considered. To test a complex device such as a microprocessor completely for every instruction in its repertoire and to apply every possible bit pattern to each instruction, the generally accepted formula for the number of test combinations is given by:

$$C = 2^{mn}$$

where n is the data word length in bits and m is the number of instructions in the instruction set of the microprocessor.

Let us now consider a device such as the 8080 microprocessor which has an 8-bit data bus and an instruction set of around 76 instructions. Hence:

$$n = 8 \quad \text{and} \quad m = 76.$$

For this device the number of test combinations that have to be applied to fully test it are:

$$c = 2^{8 \times 76} = 2^{608}.$$

We are far more used to considering numbers in the decimal system, so we will convert this number into a decimal one. Let

$$2^{608} = 10^x.$$

Hence:

$$x = \log_{10}(2^{608}) = \log_{10}(2).608 = 0.30103 \times 608 = 183.02624.$$

Therefore the number of tests required are:

$$C = 10^{183.02624}.$$

Let us assume that each test takes 1 microsecond (an optimistic value for the 8080). The number of tests will take $10^{183.02624}/10^6$ seconds, i.e. $10^{177.02426}$ seconds. There are 3.1526×10^7 seconds in a 365-day year. The tests will hence take $10^{177.02624}/3.1526 \times 10^7 = 0.3171 \times 10^{170.02624}$ years to complete.

If we started testing one device now, the tests would be complete in 3.171×10^{169} years.

The age of the Earth is reckoned to be around 4.7×10^9 years, so it is unlikely that the device could survive for even a small portion of the required test time and would undoubtedly crumble to dust well before its test time was over.

The only conclusion that can be drawn from this is that "a complex digital device, such as a microprocessor, can never be fully tested".

An implication of this is that every computer in use has never been fully tested and never can be; at best only a very limited subset of the instructions and applied bit patterns have been used to test a device functionally. The manufacturers of devices and the producers of automatic test equipment which tests LSI devices can only deal with this problem by checking the major functions with a limited set of bit patterns and assume from those tests that it will perform correctly for all applied test patterns. As the scale of the device increases, such as the progression to 16-bit microprocessors, the testing problem becomes more acute.

When an integrated circuit has been designed and sample devices fabricated, the internal layout of the device may be such as to cause an effect known as "pattern sensitivity", where a particular bit pattern may cause erroneous operation. It is unlikely that the sensitive pattern will be used in the manufacturer's tests and the effect only becomes apparent during field trials. The output line drivers of a device are usually arranged around the periphery of an integrated circuit and because of their function they generally dissipate more power than most other components in the system. Low power devices close to these may have their characteristics altered through a temperature increase due to the proximity of the line drivers. They may then perform in an uncharacteristic manner and make the device behave erratically some time after it has been powered up.

2.5 THE SYSTEM KERNEL

Certain parts of a computer system must be fault-free if the system is to operate. Collectively, these essential parts are referred to as the system *Kernel*, which incorporates:

(a) The CPU.
(b) The system clock.
(c) The Control Bus.
(d) The Address Bus.

A digital computer may be viewed as a kernel, surrounded by peripheral circuits, which must be operative before the rest of the system can be tested. Fortunately, the system kernel may be tested simply, provided facilities have been incorporated into the system which allow the data bus to be open circuited and an instruction fed into the CPU. Typically, the instruction forced into the CPU is some form of *No Operation* instruction such as a NOP or a MOV A,A instruction. The data bus may be open circuited by including slide switches in the system which disconnect the CPU from the data bus and including one or more test switches to set the states of the data bus lines left connected to the CPU, to define a single byte instruction to it. The system kernel can then be put into a FREE RUN mode of operation. The 8 slide switches, shown as S_1 in Figure 15, are used to

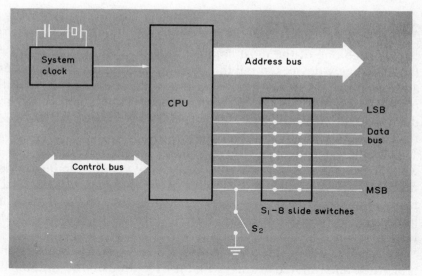

FIGURE 15   THE SYSTEM KERNEL

disconnect the CPU from the data bus. Switch S_2 is used to force the most significant line to ground, which in effect causes the CPU to fetch the hexadecimal code 7F from every possible memory location. The scheme as illustrated would be used to free-run an 8080, 8085 or Z80 microprocessor.

When all the slide switches are open circuited and S_2 is closed to ground, the CPU will execute a memory read operation to obtain its next instruction. This instruction will always be interpreted as a NOP type of instruction, which will cause the CPU to step onto the next memory address and execute a further read operation. The CPU is forced into reading a "do nothing" type instruction at every possible memory location, the effect of which is to cycle the address bus through every possible bit pattern. By examining the address bus lines, each line can be tested for a short to ground or to the positive supply rail, an open circuit, or a short to another address bus line.

Provided that the address bus lines cycle correctly, then the system kernel can be assumed to be functioning. A fault in the system clock or a faulty control bus line will almost certainly prevent this free-run mode of operation from executing correctly and will force the tester into examining the various parts associated with the system kernel. The free-run test provides a simple method of testing some of the major components of a microprocessor based system and can be applied to any type of device. The free-run mode of operation also plays a part in a testing technique called *Signature Analysis* which is discussed in Chapter 7.

2.6 CPU TESTING

Although one of the most complex, the microprocessor is one of the least likely devices to fail in a system. As we have already seen, any complex LSI device such as a microprocessor cannot be fully tested which forces manufacturers to only functionally test the devices. This typically means that each possible function has been tested using a limited set of bit patterns.

The simplest form of field testing of a CPU is to free-run the system which demonstrates that the CPU is correctly reading an instruction off the data bus, setting address patterns onto the address bus and is responding to the system clock. By checking some of the control bus lines, such as the RD/$\overline{WR}$ line, the control bus can also be partially tested.

Invariably, during field testing, the CPU is assumed to be functional and little or no isolated testing, with the exception of free-running the system, is carried out specifically on it. Almost all testing involves running some stimulus program to exercise the system, which itself assumes that the CPU is capable of executing the test programs.

A faulty CPU cannot test itself, which forces the designer of a system which has to include CPU checks to incorporate a second CPU which will simply act as a check on the first. The overall complexity of the system will clearly drastically increase with the inclusion of extra hardware to support the second CPU and will cause a corresponding decrease in the reliability of the system. Such an approach is seldom adopted directly because of the escalating problems of testing the second CPU and its associated circuitry. The idea of using a second computer system to test another is, however, a valid one, and many sophisticated test instruments used to test microprocessor based systems, themselves are microprocessor based.

2.7 ROM TESTING

The term *Read Only Memory* or ROM covers a range of device types which differ in the manner in which a program is entered into them. As the name suggests, ROMs can be read from but not written to during normal program execution. The abbreviation ROM itself is usually reserved for those devices which have the list of instructions stored within them, "written" in by the final interconnection pattern laid down during the manufacturing process. The stored bit patterns are thus fixed by this final metalisation mask and when the device is sealed the stored programs cannot be altered.

The abbreviation PROM or *Programmable Read Only Memory* is

associated with those devices which use fusible links to store the program's bit patterns. Each bit stored within the device is saved in a "cell" which consists of a single transistor. The transistor is usually a bipolar device with the fusible link in its emitter circuit. During programming, these links are either left connected or burnt out by passing a current of about 1 ampere through them. PROMs are user programmable devices and are manufactured with all the fusible links intact. The user selectively burns out the fuses to store the bit patterns of the program he desires, and in common with ROM devices, once the patterns have been stored, they cannot be altered. In the strict sense, some program modification can be achieved after programming a PROM because those links that were left intact can be burnt out by a second programming operation. This can be the case where a stored data constant is undefined during the initial programming exercise and all its associated links are left intact. Some time later, when the data constant has been decided, the PROM is reprogrammed to include the correct data constant.

EPROMs or *Electrically Programmable Read Only Memory* devices are those which can have their contents erased, usually by exposure to ultraviolet radiation of a fixed wavelength and which are electrically programmed by the user. EPROMs can thus be used to store different programs at different times by erasing and reprogramming them.

EAROMs or *Electrically Alterable Read Only Memories* are also referred to as read mostly memories and can be reprogrammed during normal use within a computer system. EAROMs have developed from a need for read/write memory devices which are non-volatile.

The type of ROM device used in a system is governed by the volume of systems to be manufactured and the type of use to which the system will be applied. The manufacturer of the device has to produce a mask to store the users bit patterns in ROM which entails a relatively high initial set-up cost. Once the mask has been made, ROM devices can be manufactured in quantity at low cost. ROMs then are predominantly used for high volume products where the mask cost can be spread over several thousand units.

Where a product is to be manufactured in quantities less than that usually associated with ROM devices, PROM devices represent the most cost effective solution.

For low quantity products and development work, EPROMs provide a flexible although relatively expensive storage medium. The vast majority of semiconductor read/write memory devices are volatile, i.e. they lose their contents when power is removed from them, while ROM type devices are non-volatile and retain their stored instructions when power is removed. Early computer memories used ferrite cores to store bit patterns as a magnetic effect and were read/write memories that were also non-volatile. Non-volatility is a desirable attribute in a computer memory system because without it, if power is removed, the system has to be reprogrammed when power is restored. ROM is non-volatile, but read/write memories are volatile and in those situations where the computer produces values and internally stores them for future use a power failure could prove disastrous or at least require resetting of the system on power-up. Certain systems, such as guided weapon control schemes, require non-volatile read/write memory as constants are updated in flight. Ferrite core memories are relatively bulky and can be replaced with EAROM type devices. In such systems, cost is a minor consideration, while size and weight problems will predominantly decide the system's components.

The vast majority of ROM type devices store fixed instructions which remain unaltered during normal program execution. This fixed nature of the information stored within a ROM provides a means of testing such devices for validity of the contents. Consider that we have a ROM device which only contains the two hexadecimal numbers 06 and 07. Adding these numbers together gives a result of OD which is then stored in the last location in the device. During test, a program is run which adds together all the ROM contents with the exception of the last location and then compares its result with the contents of the last location. If the two numbers agree, then the ROM is assumed to be correct but if they differ, then a fault is inferred in the ROM device. The stored number is called a CHECKSUM and provides a simple but effective means of testing any ROM type device where the stored information is fixed.

The CHECKSUM technique can be applied to any size of device by adding together all its contents with the exception of the last location and ignoring any numeric overflow that occurs. If, for example, the ROM contained the three hexadecimal numbers 7C, 20 and 8A, then their addition would produce the result 126 which would represent a 2-byte result. If each location within the ROM were byte wide, then only the lower 26 portion of the result would be stored as the CHECKSUM. Thus the overflow into the next significant byte of the result would be ignored by the test program to provide a result of 26 which would be then be compared to the stored checksum value. Multiple errors within the ROM could lead to a result that was an exact multiple of 256 greater than or less than the true result. This would then be seen by the test program as a correct result and the ROM, although in error, would be passed as correct. This is fortunately a remote possibility requiring at least two errors.

In those systems that use more than one ROM type device to store the programs, the last stored instruction in the first ROM is made a jump instruction to pass around the checksum location. In every ROM used, the last location is reserved for the checksum for that device and jumps are included to skip over the checksum locations.

A checksum test program is usually implemented as a program in a system test ROM which is run during the power-up sequence of a computer system. Ideally, the test ROM has to check itself as well as the rest of the system.

2.8 RAM TESTING

Random Access Memory or RAM is the name applied to memory devices that can be written to as well as read from during normal program execution. The name itself stems from early computer memories which had to be sequentially accessed by starting at location one and then stepping through memory until the desired location was reached. To distinguish between this type of memory and those in which the address was directly applied to the device to directly access the desired location, the name random access was coined. The vast majority of memory devices in use are random access in this sense, although the term has come to be associated with read/write memories specifically.

Semiconductor RAM devices are in the main volatile which results in their contents being lost when power is removed from a system. Unlike ROM devices, their contents are not invariant and thus the simple type of CHECKSUM test cannot be applied. There are essentially two types of RAM in general use: STATIC RAM stores each bit written to it in a "cell" which will remain in one state until the opposite logic state is written into it or power is removed; DYNAMIC RAM uses a simpler cell structure to increase the amount of storage within a device but whose information leaks away with time. Typically, a bit is stored in a dynamic RAM as charge on the gate-source capacitance of an MOS transistor. This charge will leak away unless periodically restored during a refresh cycle. Most dynamic RAMs have to be refreshed every 2 milliseconds or their stored information will be lost. To refresh a dynamic RAM, extra digital logic has usually to be provided in the system. To overcome this requirement, dynamic RAM devices are being manufactured with the refresh circuitry built into them. The name quasistatic has been applied to such devices because they appear to the rest of the system as if they were static type devices but are in fact dynamic RAMs.

To test RAM devices, particularly during the power-up sequence of a computer system, data has to be written into each location and then read back. If the bit pattern read back from a location agrees with that written to it, then the location is assumed to be correct. The patterns written to and read from RAM locations are referred to as CHECKERBOARD patterns because they alternate between the two logic states, analogous to the alternation between black and white squares on a draughts board. The usual patterns used are 0101 0101 (55 hex) and 1010 1010 (AA hex) which are written to and then read back from each RAM location.

CHECKERBOARD patterns are used during power-up to test RAM devices and also serve to define the upper memory address of RAM space within a system. Many computer systems are supplied as a basic model with a limited amount of RAM memory which can be expanded by the addition of extra RAM devices up to the maximum permitted by the system. The amount of RAM in a system may be from the minimum supplied with the basic model up to the maximum permitted. The test program writes a checkerboard pattern to each RAM location, starting at the lowest memory address known to be occupied by RAM and increments the test through every RAM location until one is found which does not provide the test pattern back when read. This location in a functional system will be the memory location, one above the highest RAM location, and thus defines the upper limit of RAM in the system. Typically, the total amount of user RAM available is then indicated to the operator by displaying its upper limit address on a VDU. If the displayed value is less than that expected, then a faulty RAM device can be suspected and its memory address will be displayed on the screen as a guide to its location within the system.

In many cases the supposed worst case test patterns of 55 or AA are not used and OO and FF are used in their place. While checkerboard patterns provide a simple means of initially testing RAM devices, they are by no means exhaustive. When a RAM is being tested in a suspect system, a more stringent test is to write to and read back a "walking ones" pattern For this test, the following bit sequences will be sent to and read back from every RAM location:

$$0000 \quad 0000$$
$$0000 \quad 0001$$

```
0000 0010
0000 0100
0000 1000
0001 0000
0010 0000
0100 0000
1000 0000
```

The test is meant to indicate any sensitivity of adjacent data lines to pick-up which would manifest itself as two adjacent bit positions both being set to "1" when read back from the tested location.

Within a device, lines conveying data and address information are in close proximity to each other and the internal layout of a device may be such as to cause an effect called "pattern sensitivity" to manifest itself. This effect, where a particular bit pattern causes coupling onto other data and address lines will often not be found by testing with checkerboard patterns. Fortunately for the majority of users, pattern-sensitive devices are usually discovered during development or sampled field tests of devices.

Interference can also occur between address bus lines due to a short between lines external to a RAM devices or be an internal fault. Tests can be applied to indicate any "address related" problems in a system by initially writing OO to all RAM locations and then writing FF to the first RAM location only. Every other location within the device is then checked for OO. If the test is successful, the first location is cleared to OO and FF written into the second location only and all other locations again tested for OO. If a location other than that set to FF gives a result other than OO, then an "address related" fault is assumed either in the device or on the external address bus lines. The offending bit or bits in the address bus are found by Exclusive ORing the start address of the RAM with the address which gave an incorrect read. If, for example, the base address of the tested RAM was 80 hex and a fault was found at location C8 hex, then by applying the Exclusive OR operation the offending address bus lines can be found:

$$
\begin{array}{ll}
 & A_7 \ldots A_0 \\
80 \text{ hex} & 1000\ 0000 \\
68 \text{ hex} & \underline{1100\ 1000} \ \oplus \\
 & 0100\ 1000
\end{array}
$$

Thus in this case the offending lines are A_3 and A_6. The system will then be checked for faults on these lines and/or the RAM device changed by substitution.

2.9 INPUT/OUTPUT TESTING

Input and output devices are difficult to test because there is no direct feedback connection from them back into the system. Memory devices can be viewed as a feedback loop in a system where information is sent to them over the address bus and they respond by feeding back information over the data bus. No such loop exists for the majority of I/O devices which simply receive information from the computer system for subsequent dissemination to external devices or send information into the computer system upon request from the CPU.

Unfortunately, not only are the I/O stages of a computer difficult to test adequately, but they are also an area where failure is likely from electrical overstressing of the components. In this context, the stages are assumed to include any interfacing circuits as well as the I/O port. An output line, for example, may be connected to a relatively long cable which has an associated distributed capacitance and inductance. Logic signals generally exhibit rapid transitions from one state to the other when switched, which demand that a driver connected to the line is capable of rapidly charging the line capacitance when the output state goes to a logic "1" and rapidly discharging the line when the output state switches back to a logic "0". The line driver is thus a highly stressed component and will be one of the most likely components to fail in a system.

The standard approach to I/O testing is to feed in known stimulus signals through input ports and verify them with a test program and to send known signals from the computer to output ports and have some external means of displaying and validating the data. For a general purpose computer, where the I/O port connections can be freed, a test box can be connected to input known test patterns from a set of switches and to receive output port data and display it on a set of LED indicators. The same approach can be taken for a dedicated control system providing that facilities have been included to disconnect the equipment being controlled and a test box facility is available to connect in its place.

Certain programmable I/O devices such as Zilog's PIO have an internal feedback connection between output and input. When a port is set up as an output port, the input stage is still operable which means that data latched out to the output port can also be read back into the system by treating the port as an input port.

The bit pattern written into the I/O select register defines which I/O lines will act as output lines and which as inputs. If the port is set up to be an output port only, the internal connections between the output

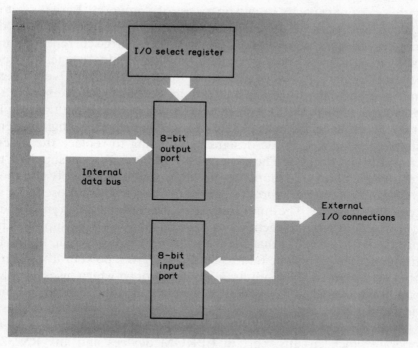

FIGURE 16   ONE PORT OF A PROGRAMMABLE I/O DEVICE

and input stages can be used to read back into the computer system any data latched to the PIO as an output port. This facility allows the device to be tested up to the external I/O lines. Caution must be exercised when using such an arrangement, to prevent any externally connected device from operating inadvertently when the test program is being run. If the device is part of a dedicated system, then the controlled equipment will have to be disconnected before the tests are applied.

2.10   SOME COMMON SYSTEM PROBLEMS

In this section we will review some of the more common faults found in computer systems; some of the failure mechanisms are common to electronic systems in general, while others relate specifically to computer systems.

2.10.1   Power supplies

Power supplies are probably the part of the system most subject to electrical overstressing and failure. A typical computer installation will use a 5-volt power supply to provide energy to the logic elements with a current capacity usually in excess of 5 amperes. Large systems will require current drive capabilities in the tens or hundreds of amperes range. Conventional designs for these large currents require bulky transformers, large rectifiers and smoothing capacitors and several regulating devices to control the final voltage. Overvoltage and overcurrent protection will usually be built into the supply to prevent a fault in the rest of the system from damaging the supply itself. To reduce the overall size of these low-voltage, high current supplies, switching mode designs are tending to replace the older, conventional designs. The essence of a switching mode supply is to rectify the mains a.c. supply to provide a rough d.c. supply and then to use the rough d.c. supply to energise a high-frequency oscillator at typically 20 kHz. The high-frequency oscillations are transformed down, smoothed and rectified to provide a low-voltage supply. The principal advantage is that the transformer used to handle the high-frequency supply will be far less bulky for a given power rating than a 50 Hz design. The result is a power supply which uses smaller components for an equivalent power rating than a conventional design would allow.

Many systems will also require power supplies at ±12 volts to operate clock circuits, certain EPROM devices and any RS232C interfaces used by the system.

A complete failure of a power supply is generally easy to detect from obvious indications such as every indicator in the system being unlit and can be verified using a *Digital Voltmeter*. A power supply may appear to fail due to a component fault in the computer which draws excessive current from the supply. If the supply is fitted with overcurrent protection, it may still supply a substantial voltage but restrict the amount of current drawn from it. In this case the system may exhibit erratic behaviour if the excess current only just exceeds the limit defined in the power supply.

Many of the components used in computer systems are critically dependent on their applied voltage levels and may only tolerate small deviations from their nominal values. A device meant to operate on a nominal 5-volt supply may only tolerate voltages in the range 4.75 to 5.25 and behave erratically outside this band. This intolerance to supply level may become pronounced in those schemes where several parts of the system are remote from each other.

Let us assume that we have a system which consists of two subsystems; system A is located in the same instrument rack as the power supply, and system B is remotely located away from the power supply unit. Both systems are connected to the power supply with standard 7-strand, 36 a.w.g. ribbon cable which has a resistance of

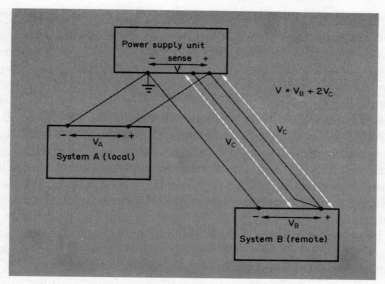

FIGURE 17   A SINGLE POWER-SUPPLY DRIVING A SPLIT SYSTEM

0.06 ohm/metre. System B is sited some 3 metres from the main instrument rack containing system A and the supply. To compensate for voltage drops along the interconnecting cables, the power supply has a "sense" facility which enables the voltage at some remote point to be set to the nominal value. The output from the power supply will be higher than this nominal value by an amount equal to the voltage drops in the supply cable and in the negative return cable. The sense input itself draws little current and the potential drop along the sense wire is assumed to be zero.

Such a scheme is illustrated in Figure 17. If the nominal voltage is set up on system B, then the potential at the power supply terminals will be:

$$V = V_b - 2V_c$$

where V_c is the voltage drop along one of the cables connecting system B to the power supply. If system B draws 1 ampere from the supply, each cable voltage drop will be 0.18 volt, giving a supply potential of 5.36 volts if system B potential is set to 5 volts. The supply potential is thus 0.36 volt above that of system B. If we assume that system A is close enough to the power supply so that the resistance of the connecting cables between them can be ignored, then the voltage applied to system A will be 5.36 volts.

Variations in the current drawn by system B will cause the supply potential to fluctuate, as it compensates for the changing cable drops and maintains V_b at the nominal 5 volts. This varying voltage level is applied to system A which could behave erratically if the potential should vary outside its acceptable limits. In practice, many systems tie the sense input and the + output terminal of the power supply together so that the supply itself maintains its output at some defined potential. System A is sufficiently close to the supply so that its applied voltage will be at the nominal set potential, but system B will have an applied potential which is not only less than the nominal value but which will also vary as the current drawn from the supply fluctuates. Under these conditions the remote system can behave erratically which would cause a tester to investigate system B, while probably overlooking the obvious point of measuring the voltage applied to it.

The only real solution to this problem is to include a second power supply unit to feed the remote system. This may only involve adding an extra regulator stage to the existing supply.

Another problem associated with power supplies, but not directly a fault in them, is that of short circuited supply decoupling capacitors. For logic circuits, a decoupling capacitor of 0.1 µF is connected across the supply rails for every two logic devices used in a system. For any reasonable size system this will involve many decoupling capacitors, distributed about the circuit boards. To provide distributed charge storage around the circuit board, tantalum capacitors are frequently used to provide reasonable capacities in small volumes. Reverse voltage spikes induced onto the supply rails can cause tantalum capacitors to become short circuited and thus cause the supply to drop out on overcurrent.

The problem of one decoupling capacitor shorting out a power supply is depicted in Figure 18. If the power supply remains on, which will be the case if it incorporates a current limiting circuit, then all the current will flow through the faulty capacitor C_2. All the decoupling capacitors are connected in parallel, which makes the isolation of one

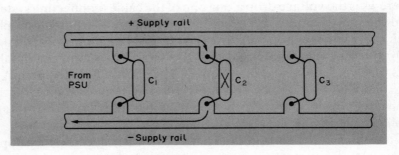

FIGURE 18   DECOUPLING CAPACITORS

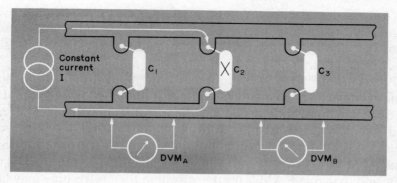

FIGURE 19   LOCATING A FAULTY CAPACITOR USING A DVM

faulty capacitor from possibly a large number of candidate components a difficult and time-consuming task.

The normal solution has been to isolate sections of the circuit board by cutting one of the supply rail printed circuit tracks and testing each section for the short circuit. Clearly this approach causes damage to the circuit board which has to be repaired after the faulty component has been isolated and changed. One alternative is to inject into the supply tracks a constant current and then to trace the current path using a sensitive *Digital Voltmeter* (DVM).

In practice this technique is difficult to implement because the supply tracks are rarely laid out in a neat and orderly fashion, which makes tracing them awkward and the voltage level will be defined by the value of the constant current, the resistance of the copper circuit board tracks and by the distance apart along the track, of the two DVM probe leads. The idealised situation is shown in Figure 19 where DVM_A will produce a reading due to the voltage drop along the circuit track while DVM_B will give no reading at all.

A second alternative discussed in Chapter 5 is to use a *Current Tracer* probe to effect the same sort of measurement but with a greater likelihood of success.

2.10.2 System clocks

The square wave oscillator which controls the operations of the microprocessor and all other clocked components in the computer system is referred to as the system clock. It governs every operation that takes place in the system and provides the basic time interval at which events can occur. Clock problems can result in a failure of the system to function at all or produce meaningless and undefined program sequences.

Clock circuits may be implemented in a variety of ways from simple

R-C (resistor-capacitor) circuits to stable, crystal controlled oscillators. With the exception of small, low cost systems, crystal oscillators are almost exclusively used to provide an accurate and stable clock frequency.

Problems can arise in microprocessor systems if the system clock runs too fast or too slow. To achieve the highest possible data throughput, a microprocessor may be run at its maximum clock rate and any deviation which tends to increase the clock frequency can cause erratic behaviour. Most microprocessors tolerate a clock which runs more slowly than the maximum specified, but if the clock is too slow and dynamic RAM is used in a system then refreshing may take place at too low a rate with a subsequent loss of information in the RAM. Both of these problems are more likely to occur when R-C clock circuits are used instead of crystal controlled circuits. However, crystal circuits can sometimes break into their third overtone (if fundamental crystals are used) oscillation modes, causing a higher than expected clock rate.

Many microprocessors require multiphase and non-overlapped clock signals which may not be TTL level compatible. A device such as Intel's 8080 microprocessor needs an external circuit such as the 8224 device to provide its two-phase clock signals, while later devices such as Intel's 8085 have all the clock circuitry fabricated within the microprocessor chip. The form the clock circuit takes then varies considerably between devices, from the direct connection of a crystal or R-C network onto the microprocessor to a separate clock circuit which provides multiphase outputs.

Fortunately, clock signals can be checked easily using conventional test equipment such as *Frequency Counters* and *Oscilloscopes*.

2.10.3 Reset circuits

A microprocessor power-up reset circuit can give rise to erroneous operation. A reset pulse that is non-existent, too short, too noisy, or too slow in transition can cause a reset out of sequence, a partial reset or no reset at all. Problems can also occur in reset circuits that are susceptible to power supply glitches which can cause a partial reset to occur and force the microprocessor into undefined behaviour. Switching a system off and then rapidly back on again can also cause the same type of behaviour by causing a failure of the supply to the device while its reset input is still maintained in a logic "1" state by a large value capacitor. The microprocessor under these conditions has not been reset, but all its internal circuits will set back up into undefined states and cause the device to behave in an unexpected way.

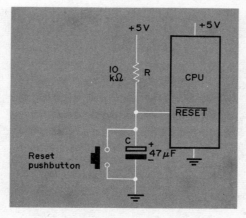

FIGURE 20   BASIC MICROPROCESSOR POWER-UP RESET CIRCUIT

The commonest form of power-up reset circuit is shown in Figure 20; when the system is switched on, the long time constant of R and C causes the reset input to still be in a logic "0" state after the supply has come up to full potential across the microprocessor. This state forces the microprocessor into a power-up reset sequence which typically takes about 20 clock pulses and internally sets the states of the microprocessors internal registers. In the Z80 device, for example, the program counter register is forced to zero, which causes the device to read memory location zero for its first program instruction. The length of time that the reset sequence takes and the operations undertaken within the device differ between microprocessors. The microprocessor stays in the power-up reset sequence until the capacitor in the reset circuit charges up sufficiently so that the device takes the potential developed across it as a logic "1" state.

To enable the device to be reset during operation, a push button may be connected across the capacitor so that it can be discharged thus forcing the reset input back to a logic "0" state.

This simple power-up reset circuit can be fooled into causing erroneous behaviour if a glitch momentarily causes a supply disconnection to the microprocessor. During the glitch time, the reset capacitor has not discharged via the resistor to any significant extent thus maintaining the reset input in a logic "1" state while the internal states of the microprocessor will be in disarray. When the supply is restored to the microprocessor, it will continue operation at some point defined by the now randomly set states of its internal registers.

The most likely cause of a supply glitch, sufficient to cause this type of behaviour, is a momentary failure of the mains supply which is often called a "brown out" to distinguish it from a failure of longer

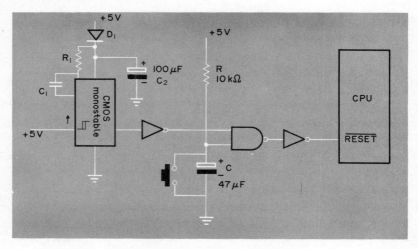

FIGURE 21   POWER FAIL DETECT CIRCUIT

duration, referred to as a "black out". "Brown outs" are random events which cannot be predicted and particularly in larger computer installations require that power-fail detection circuits are incorporated to allow the system to "gracefully decline". To accommodate this requirement, power supplies having very large storage capacitors are used to ensure that a current supply is still available for a short time after failure of the mains supply, which may also be supplemented by battery back-up systems. Clearly, in those systems where a loss of information cannot be tolerated, these measures have to be taken so that the system can have its information saved in a systematic manner before turning off.

Mains failure detection circuits typically count the mains frequency to detect the loss of several cycles and then initiate a control signal to cause an orderly switch off of the computer system. In small systems the extra circuitry involved in detecting mains failure cannot usually be justified but may be replaced with some simpler scheme which tests the supply potential in the system.

During initial switch on the reset input is held low by the normal R-C reset network. If a glitch should occur which forces the supply rail low and is then restored, the CMOS monostable circuit will be triggered, which forces a reset onto the microprocessor and initiates a reset sequence. The monostable is triggered by a rising edge which corresponds to the supply being restored and uses a Schmitt trigger input to allow for the slow risetime of the supply. The CMOS monostable is maintained in a powered up condition during the glitch duration by the diode D_1 and charge storage capacitor C_2. The circuit as it stands can only be used to cope with relatively short duration "brown outs", but could be extended to cover long duration supply interruptions by connecting a trickle charged nickel–cadmium battery in parellel with C_2.

Power-up reset circuits can be monitored by storage oscilloscopes to check their response or be manually overdriven and controlled externally for testing purposes. A permanent low voltage on the $\overline{reset}$ input to the microprocessor will keep the device locked into a reset sequence and give the appearance of a completely non-functional system. This type of fault can arise from a short to ground on the printed circuit track that connects to the $\overline{reset}$ pin, from a faulty logic device connected to the line, or from a short-circuited capacitor.

2.10.4 Interrupts

Jammed or noisy interrupt lines can cause system malfunction. A jammed interrupt input may cause the microprocessor to spend all of its time servicing a phantom interrupt, while a noisy input may cause an interrupt to the processor when none has occurred, which in turn may cause meaningless I/O transfers to take place.

The interrupt structures used by microprocessors vary considerably between devices and extend from simple fixed inputs, which have to be polled during the service routine to determine the interrupt source, up to programmable interrupt inputs with full vector capability. The actual number of interrupt inputs available on 8-bit microprocessors varies from one (the $\overline{INT}$ input) up to five on the Intel 8085 device. The majority of devices have two interrupt inputs, namely the $\overline{INT}$ input and the $\overline{NMI}$ input with the *Non-Maskable Interrupt* input having a higher internal priority over the $\overline{INT}$ input.

Most interrupt inputs to a microprocessor are edge sensitive and a change from a logic "1" state to a logic "0" state causes a flip/flop, internal to the microprocessor, to set. When the processor completes its current instruction, this flip/flop is tested and if set, forces the device into an interrupt sequence. However, some interrupt inputs such as RST6.5 input on the 8085 processor are level sensitive, which under fault conditions can cause the device to stay permanently locked into an interrupt service routine. Thus if the RST6.5 input became shorted to the positive supply rail, then the processor would jump to memory location 0034 hex and continuously execute the service routine beginning at that location.

Programmable I/O devices that can initiate interrupt requests to a microprocessor usually have their own internal interrupt latches which are set and cleared separately from the interrupt flip/flop sited in the microprocessor The latch is set by an external interrupt and cleared by some form of interrupt acknowledge from the processor when it begins to service the interrupt signal. Most microprocessors have insufficient pins to provide directly all the control signals required in a system and recourse is made to multiplexing signals through certain pins. The Z80 device is an example, where an interrupt acknowledge signal is provided by using two separate control lines which are normally used for other purposes. The $\overline{M_1}$ and $\overline{IORQ}$ control lines are taken low together as an interrupt acknowledge signal, a situation which does not occur under normal program flow.

The Zilog programmable device, the PIO, resets its internal interrupt latch at the end of its interrupt service routine by directly decoding off the data bus, the *Return from Interrupt* (RETI) instruction. This instruction consists of a 2-byte opcode ED hex followed by 4D hex both of which are latched off the data bus when read by the Z80 processor at the end of the service routine. In many systems, where a large number of devices are connected onto the data bus, buffering is used to extend the bus drive capabilities. These bidirectional bus buffers may prevent the PIO from receiving the RETI codes if they had been left in a condition to permit data transfers from the I/O device prior to the RETI codes being read from memory. Under this condition, the PIO cannot receive the codes and will not be able to clear its internal interrupt latch, a state which will prevent it from being interrupted again from an external source.

The Zilog PIO allows a user to utilise the normal I/O lines into the device as interrupt inputs; a situation that cannot be exploited with the majority of progammable I/O devices. The PIO can have any input line also act as an interrupt input and where several lines are used in this way an interrupt to the processor can be generated when either any line becomes active or only when all lines become active. The device uses an AND/OR structure to provide this flexibility, which under certain conditions can itself give rise to problems. Consider that OR logic has been programmed for four interrupt inputs, so that whenever any one of the inputs becomes a logic "1" state, an interrupt will be flagged to the microprocessor. The situation is depicted in Figure 22.

One input becomes active and sets to a logic "1" state, which causes the output from the OR gate to also set to a logic "1" state. The change in state of the output of the OR gate causes the PIO's internal interrupt latch to be set and flags the microprocessor that an interrupt is pending. The response to an interrupt signal is edge sensitive, which ensures that even if the active interrupt level persists after execution

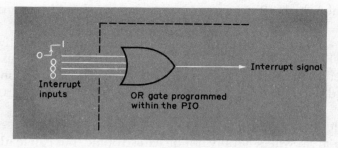

FIGURE 22   ORing Interrupt Inputs Within a Zilog PIO

of the interrupt service routine, no further interrupt signal will be generated. Unfortunately, if the first interrupt signal persists at an active logic "1" level, then, due to the OR structure programmed into the system, no further interrupt inputs can cause an interrupt to the processor. To prevent this possibly dangerous situation from occurring, the service routine must reset the active interrupt state back to its inactive logic "0" level and thus free the system to respond to any further interrupt signals that may occur.

2.10.5 Memory devices

Memory failures in a microprocessor system can cause erroneous behaviour in a number of ways. Complete devices may fail or single bits within devices may cease to function correctly. The severity of the fault depends upon in which type of device a failure occurs and its position in the memory map of the system.

Failure within a ROM type device will generally be catastrophic, because these usually contain fixed system programs that have to function correctly. If a bit failure occurs, then it may be located in an unused part of the device and have no apparent effect, although the likeliest result will be that the processor will read the corrupted word as an erroneous opcode and attempt to execute some meaningless program sequence.

EPROM failures can eventually occur due to the frequent recycling of the devices. When an EPROM is programmed, the channels of the MOS transistors contained within it are avalanched, which eventually destructively damages the devices after a limited number of reprogramming operations. Early EPROMs, such as the 1702A and some 2708 devices, can be discovered in this state having been converted into ornate tie pins.

Fusible link PROMs can also fail due to a mechanism called "growback". Early PROM devices used a fuse made from nichrome material, which after having been burnt out should have remained broken, but which could after some time reform itself into an electrically conducting link. Stored bits would then apparently change state when "growback" occurred. The early nichrome fuses were replaced by polysilicon fuses which when burnt out form an insulating silicon dioxide layer and which does not suffer from the "growback" effect.

Single bit failures in a RAM device, where it forms part of a data storage area, will generally not be catastrophic, but can lead to what are termed "soft errors". These are errors that occur when a wrong value data word is used in a computation. The result of the computation may not be as expected, but the system is unlikely to fail from it. Unless a power-up RAM test is applied from a test ROM program on switch on, these types of error are difficult to locate.

A RAM failure in the area of the system STACK will be catastrophic and cause the processor to jump to some undefined memory location when returning from a subroutine or after executing an interrupt service routine.

Faulty dynamic RAM refresh circuitry is another cause of supposed RAM failure which has to be considered when diagnosing apparent RAM failures.

2.10.6 Signal degradation

The address bus, data bus and control bus lines in a microcomputer system connect to many devices within the system. When a line is set by the microprocessor, the signal will propagate along the line from the processor to the furthest extremity of the line. Every device connected onto the line will represent a discontinuity in the characteristic impedance of the line and result in a reflection of the signal. For a short time at least the signal on the line will be subject to multiple reflections and oscillation due to the unbalanced nature of the line connections. To overcome these effects, most microprocessors allow at least one clock period for the lines to stabilise to their new values when information is placed on them. Lines that are closely packed together on a circuit board are susceptible to cross-talk from these multiple line reflections and the problems are most acute on adjacent signal lines that carry clock signals and device enable signals. Long, widely dispersed signal paths may already be taxing the timing and noise margins of the system to the limit, and an increase in humidity may be all that is required to cause signal cross-coupling between lines. To check one circuit board in a multiboard system, it may be extracted from a racking system and an extender card plugged in, to enable access to the card being investigated. The extender card itself may cause cross-coupling of signals in humid situations where critical signals lie adjacent to one another on the extender.

The display of almost any signal line used in a microcomputer system on an oscilloscope will reveal a waveshape that is certainly far from the idealised waveforms that are usually depicted as timing diagrams on a device's data sheet. The computer system is essentially a sampled system and the untidy waveforms that occur in them are usually compensated for by using the line state only at certain stable instances in time when, hopefully, all line reflections have ceased. To achieve some balancing of the transmission lines, the extreme ends may be terminated with balancing resistors, although in most systems even these are not fitted.

Chapter 3

System Testing Philosophy

Although there are several tried and tested methods of systematically checking a system for a fault, the most obvious and often ignored first step should simply to have a good look at it before connecting any test equipment. The fault may be an obvious one such as a burnt out component or the state in which the system has stopped may itself indicate the area in which the fault has occurred. Any indicators built into the system should be checked as a guide to the fault, and towards this end many manufacturers build functional go/no go indicators into modular systems as a means of identifying faulty sub-systems. Manufacturers have unfortunately been slow to implement such facilities, which can greatly reduce the down-time of a system and reduce first line maintenance to scanning the functional indicators, finding the unlit one and replacing the faulty module. Such an expedient greatly reduces the down-time of a system and reduces the fault-finding problem to that of second line maintenance where the board can be either sent back to the manufacturer for repair or serviced by the user.

The troubleshooting philosophy for microprocessor based products is fundamentally no different from that applied to standard digital designs, and in common with standard test procedure the tester should be acquainted with the system that is being tested, should understand the test equipment that will be used and have some logical procedure to follow to find the fault. The most complex and sophisticated piece of test equipment available to a tester is his own brain, which should be used to make reasoned judgements at each stage of testing a system and not simply be used to follow some blind set of rules.

Test personnel should be familiar with the system being tested and have available ALL the documentation relevant to it. This includes the circuit diagrams, any change sheets pertinent to the particular series of system under test, and where, necessary, circuit descriptions, if only in a brief form. There is nothing more frustrating than isolating a fault to one particular section of a system and then discovering that diagrams are not available, or that the diagrams that are in use do not correspond to the system under test because alterations were not amended onto the diagrams. In common with the rapid expansion of microelectronics, the tester is faced with dealing with an ever-increasing number of systems which vary considerably in detail and with which he must be familiar if he is to function satisfactorily. There is no shortcut to this learning process. This book and others like it can provide information about likely fault areas and indications as to the type of test equipment that should be applied in certain situations, but in the final analysis the tester has to test one specific system, usually with a limited range of test equipment and often under far from ideal conditions. In any practical situation then, the tester has to use his own knowledge of the system and the likely fault areas, possibly gleaned from past experience, to locate and cure a fault.

The range and scope of test equipment available specifically for microcomputer systems has developed from conventional equipment such as DVMs and oscilloscopes, which have a restricted use for computer test applications, up to special instruments such as *Logic Analysers* and *Emulators*, which have been developed specifically to cope with the particular problems associated with computer systems. Even a small range of these specific instruments represents a considerable capital outlay and the likelihood is that a test department will only have available a limited set of equipment.

A tester should be familiar with these items and appreciate their limitations and those areas in which each should be applied to minimise troubleshooting time.

The emphasis so far has been on the system hardware, but test personnel should also be familiar with software, at least those programs which are used to test a system. The test programs may be built in and run as part of a power-up sequence or be called upon demand from a test ROM. In the absence of either of these, the tester may have to load and run his own test programs from some external medium such as a cassette tape, floppy disc or plug-in test ROM. To achieve this, the tester must be familiar with the operation of the system under test to enable him to load and operate the test software successfully.

There are a number of testing problems which are unique to computing systems. Most of the control lies in the software, which makes signal flow difficult to trace. Another problem is that everything happens too quickly to be seen in real time and in most cases a microprocessor system, unlike many logic systems, cannot be stopped and manipulated. Measurements must be taken when the microprocessor is running usually at full speed, and this need to capture dynamic events reduces the effectiveness of many conventional test systems.

The bus structure of microprocessor systems imposes additional problems, because information on these busses is often unstable or meaningless due to tri-state output stages, the multiplexing of the information on lines and switching transients. As already discussed, these conditions usually cause no problems for the system itself, since it is synchronous and controlled by signals which are arranged to coincide only with stable data times. A bus structure makes it possible for many devices to be connected onto a common line, thus making the problem of finding one bad device on that line a difficult and time-consuming task.

3.1 SELF-TEST PROGRAMS

The increase in complexity of systems has spurred the need to include self-test or diagnostic software routines. These may be written into the intial part of a system ROM located in the memory space, jumped to by the program counter at power-up, or may be included on the circuit board in a separate ROM which can be switched into circuit in place of the first system ROM and jumped into by manually resetting the system.

The basic disadvantage of implementing a test ROM which is switched into circuit in place of one of the normal operating ROMs is

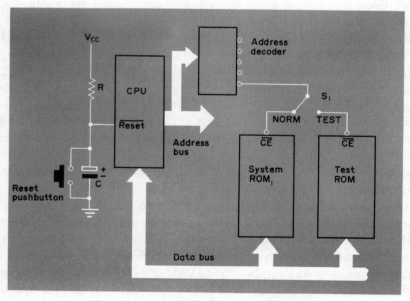

FIGURE 23   A SYSTEM TEST ROM FACILITY

that the test ROM clearly cannot carry out tests on the system ROM it replaces. If a fault is known to exist in the system and all tests perform satisfactorily, then the most likely component left to suspect is the system ROM replaced by the diagnostic test ROM. The system ROM may then be tested by substitution.

The type of tests contained within the test ROM can be a checksum ROM test, a checkerboard RAM test and I/O tests for the particular system. The different test routines can be initiated from the state of the inputs of a test input port. This test port can itself be one normally used by the system, but disconnected and a test set of switches connected instead. Any I/O tests will usually be simple routines such as sending a known squarewave pattern to an output port or sending a binary count if indicators can be fitted to the port. Alternatively, the test ROM may contain signature analysis test routines which are discussed in Chapter 7.

3.2 IS THERE REALLY A FAULT?

The complexity of systems increases as newer, more sophisticated, devices become available and as better and more powerful software is written to run on the systems. Users can often invoke apparent errors in these systems through lack of experience in operating them, because they simply pushed a wrong button which the software had not been designed to cater for. The operator of a general purpose microcomputer system has to understand not only the programming language in use but also the operating system used by the system if he is to call up and use the facilities available successfully. User errors often stem from a lack of knowledge on the operator's part about the form and syntax of calls to various parts of the operating system. A mistake in attempting to invoke a part of the operating system will often lead to a terse error message being displayed which through inexperience is frequently interpreted as a fault.

When a fault is suspected, a great deal of information can often be obtained without removing the equipment covers. Most systems have some form of input keys and displays which may be used to solicit responses by actuating keys and observing the display responses. If, for example, all indicators are unlit, then one might suspect a faulty mains switch, a broken mains lead or a power supply failure. On the other hand, if only one segment of a 7-segment display remains unlit, the fault probably lies in the display itself or is it driver circuit. Observing even the simplest of indicators may give clues as to the likely fault area. A tester should always take advantage of any built-in test routines such as power-up tests or diagnostic routines supplied by the manufacturer and which may have to be selected through an internal switch.

In all probability the fault has not been located by the "good look" approach, and the tester then has to delve deeper into the system to discover its source. Unfortunately, many people then adopt the attitude "if all else fails, look at the manual". This poor but widespread attitude makes even less sense for microprocessor based systems than for conventional ones. There may be a whole host of test procedures supplied in the manual, such as special test switches, jumpers and test indicators, just waiting to be used. Test personnel should acquaint themselves with the facilities provided before any fault develops in a system and occasionally exercise the test routines when the system is operating properly so that they are familiar with the expected responses. One integrated circuit looks very much like any other, and the tester should use the manual to familiarise himself with the location of the major components such as microprocessor, RAM, ROM, I/O, address decoders, clock and interrupt portions.

It makes sense to look first at the things that can be tested and repaired easily. Simple things are just as likely to fail as complex ones; a case in point is the power supply. This is the most failure prone part of any product and is also one of the easiest to test. An out of specification voltage can cause erratic behaviour and checking the voltage levels first may save a considerable amount of time.

A mechanical inspection can also be fruitful. Dirty printed circuit board and cable connections, broken wires and loose parts can cause excessive noise in a system (usually at low frequencies), but can usually be found either by visual inspection or touch. When applied with restraint, the nicely named "calibrated fist" test can be applied in which mechanical parts of a system are lightly struck to induce mechanical failures such as those already mentioned. Mechanically stressing a circuit board by twisting and flexing it can often help in locating poor edge connector contacts, integrated circuits loose in sockets and broken circuit board tracks. Most of those faults can be easily remedied; edge connectors can be cleaned up by rubbing them with a standard pencil eraser. These stress techniques must be applied with restraint or they may themselves induce further faults into the system.

Where possible, a test history for a system should be compiled which indicates the effects of the fault and it source. If a fault recurs, then it may be diagnosed quickly from the fault history. A fault history is particularly useful where several similar systems are in use and which will generally share the same type of fault problems.

3.3 THE LIFE-CYCLE OF AN INTEGRATED CIRCUIT

Any microcomputing system consists of a collection of *Integrated Circuits* (ICs) which vary from simple gates through to complex LSI devices such as the microprocessor. Integrated circuits in general fail through a number of known modes which mainly affect the parts of the circuit which connect with the outside world; these are the parts of any circuit which are subjected to the greatest amount of electrical stressing and which dissipate most power.

An integrated circuit is manufactured in a fabrication plant and is purchased by a product manufacturer who inserts it into a printed circuit board which is then installed into a product. The product goes into service and remains there for the rest of its useful life. Needless to say, some ICs do not lead long and healthy lives.

Unless specifically requested, ICs purchased from a manufacturer have only been batch tested, i.e. samples from a production run have been taken and tested. If a percentage pass all tests, the batch is deemed to be satisfactory and all are released.

About 2 per cent of ICs bought by product manufacturers are defective. To test incoming ICs at this point costs about 4p per device using a specially designed tester but once an IC has been loaded into a printed circuit board by a product manufacturer, the estimated cost of fault-finding and repair increases to about 40p per device.

If the defective device, mounted on a printed circuit board, is assembled into the end product, the estimated unit cost of fault finding and repair increases to about £4.

Replacing a bad IC in the field is even more expensive, at around £40.

Generally, the cost of finding a faulty IC increases by a factor of 10 at each step from the incoming component stage up to the end product in service. It clearly makes economic sense to find faulty components as early as possible. Many system manufacturers do not apply quality assurance checks on incoming components and may rely instead on equipment burn-in tests. Burn-in tests are applied at the printed circuit board level and consist of operating the equipment for extended periods of time, often at elevated temperatures, to accelerate any likely component failures.

An integrated circuit is subject to a "bathtub" life expectancy curve. Immediately following its manufacture, an IC is liable to failure from a large number of possible mechanisms such as poor hermetic sealing of its package, poor device package connections, pattern sensitivity at elevated temperatures, etc. A burn-in period of 100 hours will usually cause the failure of marginal devices, which may have passed through any cursory initial testing. If an IC survives the first 100 hours of its lifetime, then it will have a low probability of failure for typically 8 years. The constant electrical stressing of the device during its lifetime then increases the probability of failure, which gives rise to the final upward trend of its "bathtub" life expectancy curve. Overstressing, even for short periods, will accelerate this condition—a condition evidenced from EPROM failures from frequent recycling of the stored information.

Any integrated circuit consists of a header (the package) into which a silicon die or dice have been bonded. Connections are made between the pins on the package and pads on the die by ultrasonically bonding gold wires between them. The most probable form of failure in an IC is an open circuit bonding lead which has failed from mechanical vibration or has been burnt out by passing an excessive current through it. The effect on the circuit into which the failed IC was connected depends upon whether the failed bond is an input or output connection. A failed input connection forms a block in the circuit path to signal flow and can only be effectively diagnosed by noting the absence of an input signal change on some output from the device. A failed output connection will in the case of TTL type devices be seen as an uncertain logic level of approximately 1.5 volts.

Three other failure modes can occur internally to an IC. An input or output pin can short circuit to either V_{cc} or ground, a short can occur between two pins neither of which are V_{cc} or ground, or a failure can occur in the internal circuitry of the IC.

A failure in a system is most likely when a new product is first switched on at the factory, when almost anything might be wrong with it. Typical of the faults that occur after production assembly of a system are solder splashes shorting out circuit tracks following a reflow solder operation, ICs inserted the wrong way round and miswired circuits. Products that fail in the field have at least worked at some time and the type of fault commonly encountered during production can usually be eliminated with the majority of field failures attributable to component failures, tarnished or loose connections and shorts from dust build-up on a circuit board.

3.4 STRESS TESTING

Stress testing consists of exaggerating an environmental parameter beyond its normal operating limits and assessing its affect on a circuit. Three forms of stress testing can be applied to electronic equipment: mechanical stressing, thermal stressing and electrical stressing.

3.4.1 Mechanical stressing

A system may be stressed mechanically by tapping it or twisting and bending circuit boards. These types of test are often applied where an intermittent fault due to a tarnished edge connector, an IC loose in a socket or a tarnished pin on an IC causes occasional erratic behaviour of a system. The concept of stress testing is to make the fault either temporarily improve or deteriorate, either condition being of benefit in locating the source. A hairline crack in a circuit board track, which may only show up when viewed through a magnifying glass, can be made to separate further or close up by twisting the board. Twisting and bending a board in this way may cause a fault to appear and disappear, which is indicative of either a broken circuit track or an IC in a socket with tarnished or loose connections.

Tarnished contacts are easily cleaned using a reputable contact cleaner such as INHIBISOL, or by lightly abrading the offending connector or IC pins using a pencil eraser.

Care must clearly be exercised when applying mechanical stress to a system, or an overzealous application of the technique could itself create faults in a system. The "calibrated fist" and bending and twisting of boards should, however, not be disregarded as invalid weapons in a tester's armoury, and when applied with care can often indicate the type of fault in a system and its possible location.

3.4.2 Thermal stressing

Components sometimes fail from a gradual build-up of heat due to an internal fault, but which up to some temperature limit will function properly. The system containing such a faulty component will behave correctly until the device overheats. Components that exhibit this type of behaviour can usually be easily located by touching them. The condition can be eliminated temporarily by forcibly cooling the offending component using a proprietary freezer spray which can be used to pinpoint the faulty device or component. Alternatively, a device may be only marginally operating and any slight rise in temperature may cause it to fail totally or partially. The effect can be induced artificially by applying a heat gun onto it to raise its temperature sufficiently to cause failure.

Thermal stressing then consists of changing the environmental temperature of devices and components beyond their normal operating values, either to induce a fault or temporarily clear it. Cooling with a freezer spray can be used to isolate a fault more precisely than applying a heat gun, because the freezer emits a fine spray while the hot air flow from a heat gun is generally much wider. Touching each device on a circuit board can often be used to located a device that is running too hot (much hotter than others), although some devices may run much hotter than might be thought during normal operation. Devices of the same type may be dissipating different power levels because of their different frequency of use in a system, and this can rise to misleading results when attempting to assess the relative temperatures of similar devices. A faulty device can run hot enough to burn the fingers, so the technique should be applied with caution.

3.4.3 Electrical stressing

Electrical stressing consists of varying the voltages applied to a system to either determine its operational limits or to locate a marginal device. The technique must be applied with extreme caution, because severe overstressing will almost certainly cause catastrophic failure of many components. In a well-designed system, electrical stressing should never need to be applied, because the system should have been arranged such that each and every device within it receives its nominal supply potential plus or minus a tolerance figure defined on their data sheets. The most probable cause of a device becoming marginal in use, other than a newly developed device which might have become uncharacteristic in behaviour, is that the supply potential has fallen from some nominal level down to the lower value of the tolerance range of a device. This can be checked by measuring the supply potential with the system operating and adjusting its level back to the nominal value.

3.5 ISOLATION TECHNIQUES

Before any fault can be pinpointed, it has to be isolated to one area of a system. Fault-finding in this sense can be viewed as a "top-down" technique where one is initially presented with a faulty system which then has to be broken down in sub-systems, one of which then contains the fault. Depending on the size of the overall system, each sub-system may have to be broken down into smaller sub-systems before the fault is finally located. The manner in which a fault can be isolated in this sense depends upon the form and construction of the overall system. The approach taken will differ between a system that is completely contained on a single circuit board and a system which is distributed over several rack-mounted circuit cards.

Once the easy things have been tried and tested unsuccessfully, individual skills, intuition and a knowledge of the product really make a difference. Take advantage of any built-in isolation features such as selected board removal, service links and jumpers and any special test modes. These can be very useful to separate parts of a microprocessor based system to allow each portion to be diagnosed separately. Interface circuits and peripheral analogue circuits used for signal conditioning often have higher failure rates than the digital parts of a system due to higher demands on speed, power dissipation, higher working temperatures, sensitivity, accuracy, temporal and thermal drift of adjustments, external overloads and reduced safety margins. Complex, highly integrated analogue circuitry used as an interface into a microcomputer may be operated near its working limits to extract the highest possible performance from it; interference from clock lines and TTL power supply rails can cause serious noise problems.

When suspicion falls on the digital part of a system, the first thing to look for is any form of signal activity. With suitable test equipment, clock signals, bus signals, chip enable lines and control bus lines should be inspected. Absence of activity on any of these nodes indicates a possible fault. The most common failure mode for a digital IC is an open circuit input or output pin. When a pin is suspected of being open circuit, it is useful to isolate it from the rest of the circuit. A quick, non-destructive method of achieving this is to suck all the solder away from the area between the pin and its solder land on the printed circuit board using a desoldering tool or desoldering braid. The pin may then be bent to centre it in the hole and prevent it from touching the copper track at any point. This procedure is quite difficult for a plated through hole, but the pin can be pulled out to isolate it. The isolated pin may then be tested for activity or the solder land driven by shorting it to V_{cc}, or ground in the case of a suspect output pin, to check circuits further along the signal propagation path.

The techniques used to isolate digital blocks of a microprocessor based product are entirely dependent upon its electrical and mechanical architecture. If some boards can be removed and still allow the basic system to operate, then this procedure can be adopted to verify operation of parts of the system. The system may permit a free-run mode of operation which can be used to verify kernel activity, address bus activity and ROM operation.

An extender board with switches on signal and bus lines can be used to break selected signal paths between one circuit board and the rest of the system. In this way, feedback paths and stuck buses can be removed from the system. An even simpler way to open selected signal paths going through a board edge connector is to place a piece of tape over the connector edge fingers that have to be isolated.

Digital feedback loops are often difficult to troubleshoot, because errors propagate around and around. A feedback loop with a faulty output signal sends this back to the input to produce more bad outputs. Opening the feedback path prevents this, so if any controlled inputs can be generated, the signal flow from input to output can be observed. Frequently, however, it is not easy to provide this input, as many lines may need to be simultaneously controlled.

Devices can sometimes be tested dynamically *in situ* using a device called a *Digital Comparator*. This instrument is connected to the suspect device and a device similar to the one under test plugged into it; the comparator then examines the performance of the two devices and indicates which, if any, pins differ. A digital comparator can check out the majority of TTL devices while they are operating in a circuit, but does require that a spare device is available for every one that has to be tested.

The troubleshooting methods and tools used to check random logic designs are not very effective in dealing with microprocessor systems, predominantly because they cannot deal with the large number of signals that need to be viewed simultaneously if any sense is to be made of information flow within the computer. Aids such as power-up checks, diagnostic software, functional sub-system "healthy" indicators and the application of signature analysis are all viable methods of building into a system test facilities which can be used effectively to reduce the time taken to locate and correct a fault.

If any circuit boards are easy to remove and replace and known good ones are to hand, then boards can be swapped to isolate the fault to a single circuit card. When duplicates of the same board are used in a system, which can be the case in say an S100 system where several identical memory boards are in use, they can be swapped with each other to check them out. The risk involved in board swapping is that a good board can be damaged because the failure may have been an effect rather than the cause. This can cause similar problems to a replacement board as to the initial, apparently defective one. In all cases power should be removed from the system when boards are changed to prevent electrical overloads from damaging components.

Where an identical system is available, functional comparisons can be made to try and locate the faulty area. This is useful when it is not clear that an actual fault exists—it may be a product idiosyncrasy or design limitation.

If a device in a socket is suspect, try tapping it first to eliminate bad contacts before attempting to replace it. One of the last devices to suspect is the microprocessor itself, but it is often the first to be replaced. The failure rate of microprocessors is very low; however, they are complex devices and their operation is difficult to verify. This is also true of the LSI chips used in microcomputer systems. If a device is a sampled version or an early model, a phenomenon known as "uncharacteristic behaviour" may occur where a device fabricated near its specification limits occasionally drifts out of them and performs erratically. This type of fault is extremely difficult to locate, because the malfunction may only occur for certain operations and then only intermittently.

3.6 THE TROUBLESHOOTING TREE

Fault location can be greatly simplified if a well-thought-out sequence of tests is carried out. Many of these tests may be simple visual checks which can give meaningful information about the possible fault area.

A troubleshooting tree for a system resembles a flowchart and is designed to narrow down the possible fault location as rapidly as possible. Unlike an ideal flowchart, the detail contained in any one box in the tree diagram may vary considerably from a simple visual check to a self-contained, complete test procedure for part of a system. Troubleshooting trees can be decided for any system (if not supplied by the manufacturer), particularly after some experience has been gained of the system and fault-finding on it. After some time in the field, a system may be retrofitted with test features from experience gained of its more common and likely faults. These retrofitted facilities may just consist of isolating switches and connections to which test equipment can be readily fitted. Troubleshooting trees should be modified to include such facilities and to incorporate any additions made to the system during its working life.

A troubleshooting tree is followed through until the fault is located, although it must be emphasized that, in common with any other fault-finding procedure, the tester is considered to be exercising judgement as to the possible fault cause and not just blindly following the tree paths. Troubleshooting, particularly on complex equipment, is often a game of probabilities where judgement has to be used to decide the most likely cause before proceeding. Certain faults, such as a lead bond failure in an IC, may be inferred as the most probable cause of a failure, which then has to be followed by a positive decision to replace that IC to check whether the diagnosis was correct. The initial part of a troubleshooting tree for a small system which has free-run facilities incorporated can be as depicted in Figure 24.

Each diamond decision box in a troubleshooting tree asks a question about system activity, beginning with very simple questions such as whether any indicators are lit. As one proceeds through the tree,

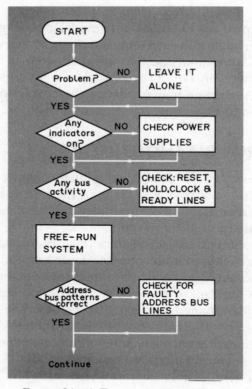

FIGURE 24   A TROUBLESHOOTING TREE

the methods and equipment used to answer the questions get progressively more difficult to apply and involve a higher level of sophistication from the test instruments. If, for example, the answer to the box, "Any indicators on?" was No, then the tree indicates that the power supplies should be checked, which only involves connecting a DVM across them and checking their potentials. To answer the question "Address bus activity correct?" fully will involve the tester in using more complex test equipment than a DVM. A free-run test usually involves forcing some form of No-OPeration type instruction into the microprocessor, so that it recurrently cycles through every possible address on its address bus. This means that starting at the most significant address line, the frequency of the information on lower address lines doubles so that a frequency counter can be used to decide if the address bus activity is correct or not. If, for example, two lines give the same frequency, then a short between them will be suspected.

As shown in Figure 25 for the three lowest address bus lines, in the free-run mode they behave as if they are the outputs of a binary

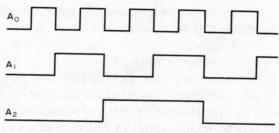

FIGURE 25   FREE-RUN ADDRESS BUS ACTIVITY

counter, with each succeeding line giving a pulse rate equal to half the previous output line. There are alternative means of deciding on correct bus activity which are discussed in Chapter 7 on Signature Analysis.

A troubleshooting tree provides a method of working systematically through a system for fault-finding, starting with major faults and ending at component or lead problems. By following the tree, an indication of the possible fault area is given and the type of tests that have to be applied are indicated. Coincidental with a troubleshooting tree, there should be provided a description, albeit brief, about the type of tests to apply and where in the system this should be done. If this information is not provided, then the tester may be faced with arming himself with several data books to determine which pin of a device contains the signal that needs to be checked.

Servicing information should be sufficiently explicit so that the tester does not need to consult other documents to find the information he needs. Unfortunately, most manufacturers do not provide this information in the correct format to ease the load on any fault-finder. Often the fault-finder himself has to generate his own troubleshooting tree diagram and also delve through the myriad of data books and sheets to discover the relevant information that should be included on it. Historically, computer system manufacturers have provided systems, but little in the way of service manuals and fault-finding charts, which are usually produced almost as an afterthought by demand from system users.

To counteract the lethargic attitude of manufacturers, several large corporations have decided their own test philosophy and force either the equipment manufacturer or a sub-contractor to supply testing information to their specifications. The Royal Navy have implemented such a scheme to minimise the down-time of their systems in the same way as the British Steel Corporation, who insist upon *Functional Systems Documentation* (FSD) to supplement any system, so that it may be rapidly tested and repaired. The lethargy of manufacturers is somewhat understandable when one considers that to write a complete test manual for a system is time-consuming, tedious, and often difficult because the system was not initially designed with fault-finding in mind. The user himself is unaware of this lack of information until the system fails and he is faced with repairing it. Both manufacturers and users alike are becoming aware of the problems involved in testing complex, computer based systems and are beginning to include built-in test facilities such as free-run checks as well as diagnostic software routines which can be updated as the system is itself upgraded.

Chapter 4

The Use of Conventional Test Equipment

Conventional test equipment is taken to be those instruments that have historically been used to test electrical and electronic circuits prior to the introduction of the microprocessor in 1970. Conventional equipment in this sense includes *Oscilloscopes, Digital Voltmeters* and *Frequency Counters*. While these instruments are in widespread use for general testing of systems, they have only a limited application to computer systems, where information is represented in a parallel format on a large number of lines simultaneously. The structure of a computer system, however, does not completely preclude their use, and there are many instances where conventional test equipment can play an active role. Several instances where conventional test equipment can be used to check out part of a computer system have already been discussed; checking power supply potentials and testing address bus line frequencies in a free-run mode are typical examples of cases where such instruments can be exploited.

The majority of test departments will carry such items of equipment, and all test personnel should be thoroughly acquainted with their operation and uses, although they should be treated as only a limited part of a tester's armoury if he has to deal in any significant way with digital computer systems and should be augmented with more specialised items specifically designed to handle the complex data formats used in computer systems.

4.1 MULTIMETERS

A multimeter consists of an instrument that can measure direct current and voltages, alternating voltages and often currents, and resistance values. The older and simpler forms of multimeter are essentially an ammeter which is configured in a circuit using resistors, rectifiers and batteries to provide the required measurements. The basic measurement in all cases is direct current, displayed on a moving coil ammeter. A typical instrument will have a 50 microampere full-scale deflection ammeter as it display device, which is often quoted as a sensitivity factor of 20,000 ohms/volt. Thus on a 3-volt full-scale range, say, the total resistance in circuit, including the resistance of the ammeter itself, will be 60,000 ohms.

On low-voltage ranges, the relatively low resistance of these instruments may load the circuit being measured sufficiently to unbalance it and give rise to a misleading reading. The newer *Digital Voltmeters* (DVM) measure circuit potential using high input resistances which do not load the measurement point to any appreciable extent. The basic measurement is thus voltage rather than current, which can only be measured by introducing a resistance in series with the circuit being measured, and measuring the potential developed across it. Typical input impedances of DVMs are in excess of 10 megohm when measuring direct voltages. Many of the DVMs in current use are based on an *Analogue-to-Digital Converter* and the display of the measured parameter will be on an LED or LCD panel display.

The primary use of a DVM, when testing digital systems, is the measurement of power supply rails and supply currents. A standard $3\frac{1}{2}$ digit DVM can be used on its 00.00 to $\pm$19.99 volts range, for example, to measure the 5-volt supply voltage associated with the majority of logic circuits in use. The converters used in DVMs have sampling rates of around 5 per second, so they cannot be used to detect any rapid voltage surges that may occur in a system due to a fault. The same will be true when they are used to monitor a supply current and they will only display an average value.

Multimeters are also used to find shorts on power supply lines (with the system powered down) due to such faults as short-circuited tantalum capacitors. Having established that a short exists, the multimeter should then play no further part in locating the fault, unless the tester is prepared to unsolder capacitors, one at a time, until the fault component is located. In the absence of any other test equipment, this is probably the only possibility open to the tester, although, as explained in Chapter 5, there are far quicker and more efficient techniques to locate this type of troublesome fault.

Many of the newer DVMs include a continuity test facility which provides an audible tone when continuity between two test points exists. This type of feature is often useful when testing cables and circuit board tracks for suspected breaks.

4.2 FREQUENCY COUNTERS

A *Frequency Counter* consists of a gated counter in which the input signal is allowed to update the counter for a period determined by an accurate and stable oscillator. If, for example, the oscillator circuit is such that it enables the input gate for a 1-second period, the count left in after the gating time has elapsed and assuming that the counter has been initialised to zero will be a direct measure of the average pulse repetition frequency of the input signal. The counter will have a limited bit length, and to accommodate a wide input frequency range the gating period is usually selectable. Thus if a high frequency is to be measured, the gate period is made short, while for a very low-frequency input the gating period will be correspondingly much longer.

A frequency counter will be used to measure the computer clock frequency which defines every operation carried out by the computer and also any other derived frequencies used in a system. The frequencies used to control the line and frame rates of a *Cathode Ray Tube* (CRT) display may be derived from the basic system clock through a frequency divider chain, and these may be tested using the frequency counter. When information is communicated over serial transmission lines, the baud rate in many systems is defined by a frequency derived from the basic system clock and this again may be tested using a frequency counter.

The majority of computer systems use crystal controlled clock circuits which rarely drift off frequency sufficiently to cause any serious system problems, so that a frequency counter is rarely needed when testing computer systems. Frequency counters generally can be used to measure time periods as well as frequency, and this feature may be used to check the length of a control signal sent out from an output port as a simple means of verifying that the program used to generate the output pulse has been correctly timed.

Some instruments, such as the SOLATRON LOCATOR, incorporate both a DVM and frequency/period counter within one piece of equipment that also provides other extensive digital testing facilities.

4.3 THE OSCILLOSCOPE

An *Oscilloscope* allows a user to view signals in the time domain as a trace on a CRT. A typical dual-channel oscilloscope permits up to two waveforms to be viewed simultaneously on the screen, and the spread in time of the display can be controlled by an adjustable time-base oscillator, built into the instrument. A basic instrument in the form described is useful to capture and display repetitive waveforms regardless of their precise waveshape. A trigger circuit in the oscilloscope is set to cause the timebase to sweep across the screen when a preset voltage value has been exceeded by the input signal and the direction of the trigger, i.e. either a positive or a negative going signal can be selected. Thus for a repetitive input signal, once the oscilloscope has been set to trigger at the desired point on the input waveform and the timebase sweep rate has been selected, an apparently stationary display appears on the CRT screen where succeeding waveforms reinforce each other. The persistence of the display phosphors used on the CRT faceplate are long enough so that a successive input signal causes reinforcement of any previous one before the phosphor decays.

Where non-repetitive waveforms have to be captured and displayed, storage oscilloscopes are used which when triggered will acquire an input waveform and display it until either it naturally decays or it is erased prior to capturing another one. The older versions of storage oscilloscopes used a special faceplate to store charge which would eventually decay through leakage. Later storage oscilloscopes digitise the input signal and store the values in RAM. These are then converted back into analogue values and fed into a normal CRT display for viewing. The digital values stored in RAM do not decay unless the equipment is switched off and thus can be viewed indefinitely. The major disadvantage of such instruments is the limited size of RAM available, which dictates the number of levels that the input signal can be quantised to and the total number of points on a waveform that can be stored. This limitation becomes less of a problem as larger and larger semiconductor RAM devices become available.

The problem in a computer system is that during execution of a program no signals are usually regularly repetitive, and if viewed on an oscilloscope appear as almost random binary signals that change from one logic state to the other at irregular intervals. The oscilloscope cannot be triggered from such waveforms because of the irregular state changes and some form of synchronised signal has to be used, such as the system clock or some derivative from it. A dual-channel oscilloscope can only display two waveforms simultaneously, and hence is practically useless as a recording instrument if, say, the data bus needs to be analysed. The oscilloscope can be used to indicate that activity is taking place on each data bus and address bus line, but the limited amount of information that can be displayed at any one time precludes any sensible decision being made about its meaning.

Under certain conditions the oscilloscope can be used to test parts of a computer system. In Section 2.5 the free-run mode of testing the system kernel was described; an oscilloscope could be used in conjunction with this test to measure the pulse repetition frequencies on all the address bus lines to ascertain that they are all fault-free. The free-run test forces repetitive waveforms on all the address lines and an oscilloscope can be triggered off by any of them. Simple test programs can be written which repetitively cycle other parts of a system such as an output port.

	Z80 MNEMONICS		*8085 MNEMONICS*
	LD A,OOH		MVI A,OOH
LOOP:	OUT(01),A	LOOP:	OUT 01
	INC A		INR A
	JP LOOP		JMP LOOP

This simple program, given in Zilog Z80 and Intel 8080 or 8085 mnemonic codes, initially clears the A register in the CPU to zero and then latches that value out to an output port at I/O address 01. The contents of the A register are then incremented and the program jumps back to send this new value out to the same output port. The program is a never ending loop which causes the data latched to the output port to cycle through the sequence from 00 to FF, reset back to 00 and continue counting again. By connecting an oscilloscope to each output line, all 8 lines can be tested.

When troubleshooting digital circuits, many gates often have to be checked. The major problem when taking measurements on gates is that most possess two or more inputs and an output which ideally need to be monitored simultaneously. The standard dual-channel oscilloscope can only display two of the signals at any one time, thus making gate tests difficult and time-consuming. Expanders can be used to split each Y input so that two signals may be displayed per channel, thus allowing four signals to be displayed at once, but to achieve this each input channel has to be switched from one signal to the other. The problem with this type of system is that one may be looking for an event which may only persist for say 10 nanoseconds and if it is to be displayed successfully the channel switch must alternate between its two inputs at a rate in excess of 200 Mhz. Circuits designed to operate at these frequencies, and which are capable of handling the voltage levels of logic signals, are critical to layout and generally expensive. Some oscilloscopes allow the triggering waveform to be displayed along with the two normal Y channel inputs, thus permitting three waveforms to be viewed together.

Particularly during the design and initial development phase of a system, a circuit may be designed where two inputs are expected to arrive at the inputs to a gate simultaneously, but due to slight differences in the propagation delays in the circuits that provide the input signals one arrives slightly ahead of the other. Such a condition is called a "race condition" and can give rise to the type of problem indicated in Figure 26. The two input waveforms to the NAND gate

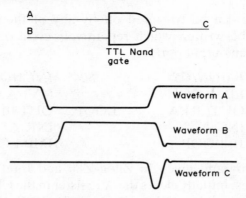

FIGURE 26   GLITCH GENERATION DUE TO A RACE CONDITION

suffer from this "race condition" and one alters before the other, producing a spike or glitch in the output. Depending upon the type of gate used, this glitch may only be several nanoseconds wide, but it may be sufficient to cause the circuits to which it is connected to trigger and give rise to erroneous behaviour. An oscilloscope which can display three waveforms simultaneously can be used to detect this type of problem which is difficult to analyse by any other means.

A similar type of problem can occur in a system, due primarily to initial poor design, but which, due to the manner in which the faulty signals are used, does not show up as a fault. If at some later stage the system is extended and use is then made of the signals which contain glitches, the system may behave erratically.

A synchronous 3-bit counter is shown in Figure 27, using J-K flip/flops, such as the 7470 TTL device. The circuit is required to count to 4 and then reset to zero to begin counting again. An asynchronous reset circuit has been used which, when the output pattern for 5 is detected, causes the counter to reset to zero. It must be emphasized that the circuit given is an example of bad design and is only given here to indicate the type of problem that can occur if circuits are not properly designed at the outset. The glitch that occurs on the Q_A output only persists for a time, determined by the propagation delays through the NAND gate in the reset circuit and through the flip/flops when resetting to zero from their reset inputs. For standard TTL parts, this glitch will be less than 68 nanoseconds wide, which if the clock period is 1 second will represent only 0.0000057 per cent of one clock period. The outputs from the circuit can, for example, be used to display a binary count on LED indicators which will not respond to the glitch and only display the counts 0 through 4 in binary. If at some later stage, say, the counter outputs are used to provide inputs to further processing logic, the glitch may cause the logic to malfunction.

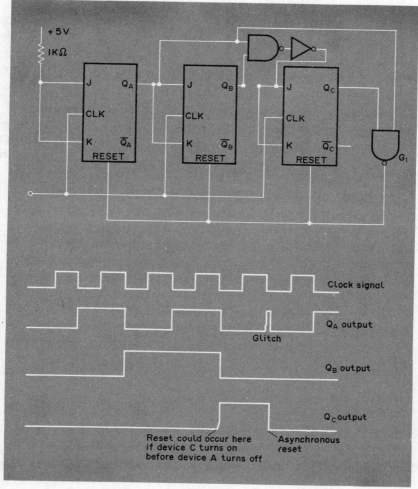

FIGURE 27   CIRCUIT GLITCH INDUCED THROUGH POOR DESIGN

A glitch of such short duration in a set of signals which change state at a much lower rate is very difficult to detect on an oscilloscope, which would have to be triggered from one of the low repetition rate signals. To solve this type of timing problem, special purpose analysers have been developed which can highlight such short duration asynchronous signals and are described in Chapter 6 on Logic Analysers.

4.4 THE LIMITATIONS OF CONVENTIONAL TEST EQUIPMENT

Conventional test equipment, although useful in detecting several of the more common problems that occur in electronic equipment, can only play a limited role when troubleshooting bus structured systems, such as microcomputers. The parallel nature of information on many lines simultaneously and the rapid rate at which the information changes prevent the usual types of instruments found in most test departments from playing any major role when analysing problems on computer systems. The explosive growth of microprocessor based systems has forced the manufacturers of test equipment to develop a range of equipment, specifically designed to locate problems in such systems. The test equipment has not developed overnight, and ranges from simple hand-held tools up to sophisticated analysers which are still being updated and improved.

Chapter 5

Hand-held Tools

Probably the earliest items of test equipment specifically designed to troubleshoot digital circuits were the logic probe, the logic pulser, the current tracer and the logic comparator. With the exception of the logic comparator, these instruments are used either to excite or to check individual nodes in a logic system and can provide valuable assistance in deciding the logic state of a node or if it is being subjected to logic activity. These hand-held tools may either be used separately when testing a system or in combination to carry out stimulus-response tests. Hand-held tools derive their power supply from the System Under Test (SUT) and are characteristically physically small and hence highly portable items.

To augment the logic probe, pulser and current tracer, there are also available logic clips and comparators. Both of these instruments are used to test functionally one complete integrated circuit while operating in a system. A logic clip is clamped onto the IC to be tested in circuit, and it derives its power from the IC's power rail pin connections. Logic circuitry inside the clip determines the polarity of the supply pins and LED indicators on top of the clip indicate the logic state of each pin on the IC. A logic clip can only test one logic family such as TTL and then only a limited range of that family because of the wide range of supply rail connections and, in particular, the large number of IC packages used. A clip may, for example, be capable of checking only 14- and 16-pin Dual in Line (DIL) packages and within that restriction only a limited number of different devices. The logic contained within a clip is usually restricted so that rapid pulse activity of any particular pin of the device under test cannot be seen on the LED indicating the state of that pin. The majority of restrictions imposed by a logic clip are overcome by the logic comparator, which takes the signals from the device under test through a passive clip and over a flying lead. An IC, identical to the one being tested, is plugged into the comparator and any differences between the tested device and this is displayed on a set of LEDs. Both devices are operated in

73

tandem, but the outputs of the device in the comparator are only used within the comparator itself to derive correct/faulty signals for subsequent display. Logic comparators are typically fitted with a personality card for the particular device being tested, which sets up the instrument and informs it of power supply, input and output pins. It is far more flexible than the logic clip and can be used to test the majority of a logic family by incorporating a number of different size sockets, into which the comparison devices are connected, and a set of flying leads with different size DIL package connectors which can be plugged into the comparator.

5.1 LOGIC PROBES

A logic probe monitors the in-circuit activity of a single point in a system and by means of a number of indicators informs its user whether the tested point is at a logic "1" state, a logic "0" state or is floating at a bad level. The majority of probes will also indicate pulsed activity of the point by flashing one of its display indicators. Separate indicators may be used to indicate the logic polarity of the point being monitored, or a single indicator may be used whose intensity varies from bright when the node is at a logic "1" state to off for a logic "0" state. If the node is floating at a bad logic level, the single indicator will be dimly lit.

The voltage levels that define the logic state of a signal are not fixed potentials, but a band of voltages which provides the digital devices with a degree of immunity to noise.

The nominal operating voltage for *Transistor-Transistor Logic* (TTL) is 5 volts. The band of potentials that represents the two logic states is shown in Figure 28, referred to the input of a device and to its

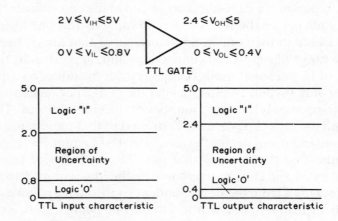

FIGURE 28   TTL INPUT/OUTPUT VOLTAGE CHARACTERISTICS

output. When a TTL device has its output state at logic "0", the actual potential developed across the bipolar transistor that connects the output node to ground depends upon the amount of current that it is sinking to ground. The input of any gate connected to the output will pass a current through the output transistor and the total current will be the sum of all the input currents that connect to the one output node. For TTL devices, the maximum guaranteed potential developed across the output transistor is 0.4 volt, provided that the number of inputs that the output node feeds into is less than or equal to a predefined limit. This is called the fan-out of the gate, and for standard TTL devices a fan-out of 10 is typical, which means that a single TTL gate output can be connected to up to 10 TTL gate inputs.

A safety margin has been designed into the system by guaranteeing that a standard TTL input will accept a potential up to 0.8 volt as a logic "0" level. Similarly, provided the design rules are kept, the minimum logic "1" output from a gate is 2.4 volts, while an input will accept any potential above 2 volts as a logic "1" level.

For the TTL gate shown in Figure 28, the following potentials can be defined:

V_{IH}— An input voltage within the range used to represent a logic "1" state; usually the minimum value is specified, i.e. 2 volts for standard TTL devices.

V_{IL}— An input voltage within the range used to represent a logic "0" state; usually the maximum value is specified, i.e. 0.8 volt.

V_{OH}—The voltage at an output terminal when the input conditions establish a high level at the output; usually the minimum value is specified, i.e. 2.4 volts.

V_{OL}—The voltage at an output terminal when the input conditions establish a low level at the output; usually the maximum value is specified, i.e. 0.4 volt.

The differences between the guaranteed input and output high and low levels determine the ability of the gate to reject noise induced into the system.

The effect of noise can be illustrated by the simple two-gate scheme shown in Figure 29. Consider that the output of gate G_1 is at its lowest guaranteed high level potential V_{OH}. This will place the output

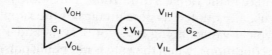

FIGURE 29   TWO TTL GATES CONNECTED BY A NOISY LINE

potential at 2.4 volts. A noise source, represented by the ideal voltage generator V_N in Figure 29, is in series with the output from gate G_1, making the potential at the input to gate G_2 equal to the sum of V_{OH} and V_N. Therefore

$$V_{IH} = V_{OH} \pm V_N.$$

V_{IH} must be at least 2 volts for guaranteed operation of gate G_2, so that if V_N is greater than -0.4 volt the input potential to G_2 will fall below the guaranteed input minimum and may result in erratic behaviour of G_2. The value of V_N that causes this unwanted condition is called the noise margin, and in the case of TTL is 0.4 volt. Expressed another way, TTL devices can tolerate induced noise levels of 0.4 volt or less before the conditions exceed the guaranteed operating parameters. Similarly, if the output G_1 is in its low state, then the input potential to G_2 is given by:

$$V_{IL} = V_{OL} \pm V_N.$$

If V_{OL} is at its maximum value of 0.4 volt and V_N exceeds 0.4 volt, then V_{IL} will exceed its guaranteed maximum value of 0.8 volt and the device will be operating outside its guaranteed limits. The noise margin in the low logic state is again 0.4 volt. TTL devices can hence tolerate noise levels up to 0.4 volt in either logic state before the conditions exceed their operating parameters.

Between the maximum logic "0" potential and the minimum logic "1" level there exists a band of voltages for which the behaviour of the device is unspecified. Usually, when switching from one logic state to the other, an input or output will rapidly transit through this region of uncertainty. If, for example, the internal bonding lead connecting an IC to an output pin becomes disconnected, any inputs connected to that terminal will cause it to float to a potential within this region of uncertainty and may in an extreme case cause gates to oscillate.

A logic probe used to investigate TTL gates has to be able to discriminate between the three possible states of logic, "0", "1", and a potential within the region of uncertainty. To be of general use, a logic probe should also be capable of testing *Complementary Metal Oxide Semiconductor* (CMOS) logic devices which have different logic thresholds from TTL devices. Unlike TTL, which is intended to operate from a fixed 5-volt supply, CMOS devices can operate on a supply anywhere in the range 3 to 18 volts. Threshold levels and noise margins are usually quoted as percentages of the supply rail voltage V_{DD}, with typical values being a V_{IL} of $0.3V_{DD}$ and V_{IH} of $0.7V_{DD}$. Under static conditions, when the gate is not being pulsed, the output thresholds are almost ideal at 50 millivolts for V_{OL} and 50 millivolts below V_{DD} for V_{OH}. The typical value of noise margin quoted for CMOS devices is 45 per cent of V_{DD}, which places it far higher than the value for TTL and is the primary reason why designers tend to choose CMOS logic for use in noisy industrial situations. The output impedance of a CMOS gate is, however, up to ten times larger than a TTL device, which makes them more susceptible to induced noise, although their larger noise immunity is sufficient to make this a minor problem in most systems. CMOS devices are often selected because of their lower power consumption compared to TTL devices, a point of particular importance where the equipment is to be operated from batteries. Logic devices are used in dynamic circuits where the system states are being switched frequently. The power consumption of CMOS devices increases rapidly at higher switching frequencies and will exceed that of an equivalent low-power Schottky TTL gate somewhere between switching rates of 500 kHz and 2 MHz. This should clearly be borne in mind when designing low-power systems which involve high switching rates, where CMOS may be at a disadvantage compared to other logic families.

A logic probe for general use must be capable of being used on both TTL and CMOS devices, although many circuits that employ CMOS gates still maintain a supply rail potential of 5 volts. The probe has to display both HIGH and LOW logic states and bad levels, including open-circuit outputs and open collector outputs without pull-up resistors connected. Besides the display of static states, a probe should also be able to indicate dynamic nodal behaviour to the user.

A probe should be capable of indicating not only the presence of short duration pulses with a high repetition rate but also the passage of isolated short duration pulses. Both of these dynamic signals are catered for by pulse stretching any short duration signal to say 50 to 60 milliseconds, which is a sufficient time period to allow a user to detect a change in indicator state from "off" to "on" and back again.

Logic probes vary in performance from relatively simple devices up to sophisticated instruments that contain integrated circuits specially developed for the application. Figure 30 illustrates a circuit for a logic probe that will give an adequate performance when testing TTL devices.

The supply for the probe is derived via a flying lead from the system under test. A 1N4001 diode is included in the supply rail to prevent damage to the probe if inadvertently the supply leads for 5 volts and ground are connected the wrong way round. The probe input is biased, using resistor values that will not significantly load the node being tested to about 1.6 volts which lies in the region of uncertainty for TTL devices. The probe input connects into a dual CMOS

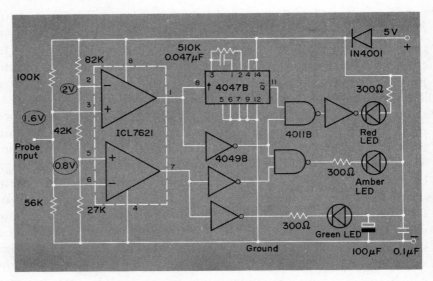

FIGURE 30   A SIMPLE LOGIC PROBE

operational amplifier which compares the potential on the probe
input against fixed voltages derived from a resistor chain.

The upper op-amp output will change from a logic "0" state to a
logic "1" state when the potential on the input exceeds 2 volts. The
lower op-amp output will change from a logic "0" to a logic "1" state
when the potential on the probe input falls below 0.8 volt. With the
probe input left floating, both op-amp outputs will be at a logic "0"
level, which causes the amber LED to light to indicate a "bad" logic
level. If the input potential rises above 2 volts, the red LED will light
and the amber LED will extinguish to indicate a logic "1" level. If the
probe input falls below 0.8 volt, the green LED only will light,
indicating a logic "0" level. Any pulse activity on the node being
probed will cause the output from the upper op-amp to change from
one logic state to the other and then revert to its original state. The
positive going edge from this device triggers the 4047B monostable
for approximately 60 milliseconds and flashes the red LED. The
circuit can hence be used to indicate static logic levels of high, low and
floating and also indicate pulse activity on a node. CMOS devices
have been used throughout to ensure that only a limited current of
typically 12 milliamperes is drawn from the test circuit by the complete probe assembly. The complete assembly may be built into a
commercially available housing measuring $30 \times 20 \times 140$ millimetres.
The primary drawback of the circuit is the bandwidth limit of 500 kHz
imposed by the ICL7621 dual op-amp which will prevent the probe
from detecting narrow glitches in the tens or hundreds of nanoseconds
region.

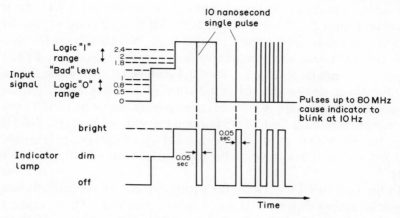

FIGURE 31   HEWLETT PACKARD 545A LOGIC PROBE TTL RESPONSES

5.1.1 Commercial logic probes

Several manufacturers produce logic probes which are capable of detecting single pulses as narrow as 10 nanoseconds and indicating pulse repetition rates up to 80 MHz. A probe may be switchable, so that it can be used on a variety of logic families such as TTL, DTL, RTL, HTL, MOS, CMOS and HiNIL. An example of a logic probe with these facilities is the Hewlett Packard model 545A. This probe stretches single pulses of 10 nanoseconds or greater to 50 milliseconds for display on an incandescent lamp housed in the tip. The intensity of this single indicator informs the user of the logic level as either high, low or "bad".

Figures 31 and 32 illustrate the Hewlett Packard 545A logic probe

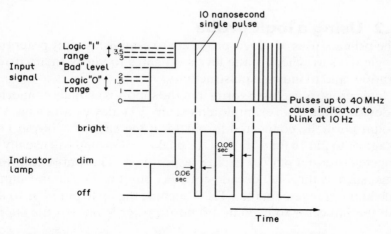

FIGURE 32   HEWLETT PACKARD 545A LOGIC PROBE CMOS RESPONSES (5-VOLT SUPPLY)

indicator responses to various input signal conditions for TTL devices operating on a 5-volt supply and for CMOS operating at the same supply potential. The logic "1" and logic "0" threshold levels are within a small band of voltages around a nominal value. For TTL, the logic "1" threshold level is 2 volts +0.4 volt, −0.2 volt, while the logic "0" threshold level is 0.8 volt +0.2 volt, −0.3 volt. CMOS, although shown for a 5-volt supply voltage, can operate between 3 and 18 volts; the supply potential for CMOS is given the symbol V_{DD} and the 545A logic thresholds are defined in terms of V_{DD}. The logic "1" threshold is $0.7 \times V_{DD} \pm 0.5$ volt, while the logic "0" threshold is $0.3\ V_{DD} \pm 0.5$ volt.

In the case of TTL devices, the probe can detect and indicate single negative pulses down to 10 nanoseconds in width. These cause the indicator housed in the probe tip to go out for 50 milliseconds or it can detect single positive pulses down to 10 nanoseconds in width, which will cause the indicator to become bright for 50 milliseconds. Pulse trains of up to 80 MHz rate can also be displayed as blinking at 10 Hz. For CMOS devices, the maximum pulse repetition rate that can be displayed is 40 MHz, and single pulses down to 10 nanoseconds in width are pulse stretched to 60 milliseconds.

The 545A probe also has a memory feature which causes an LED mounted on the probe body to stay on every time the tip detects a change in logic state. A button on the probe can be used to clear the display, which will relight at the next detected input state change. This feature is useful for capturing one or more pulses when it is inconvenient to look at the indicator in the tip; to detect pulses that occur infrequently; or when the point in the system under test is hard to see.

5.1.2 Using a logic probe

The primary uses of a logic probe are to establish supply potentials on logic devices, check static levels on gates to verify their correct operation, and to indicate pulse activity.

A logic probe is useful to verify that the supply is actually connected to a device. Take as an example almost any TTL device which has V_{CC} (5 volts) normally connected to pin 14 in the case of a 14-pin DIL package or to pin 16 for a 16-pin DIL package. Ground will usually be connected to either pin 7 or to pin 8 respectively. There are exceptions to this, such as the 7490 decade-counter, but it is the general supply connection arrangement for TTL. Touching the tip to pin 14 or 16 will cause the indicator to become bright if power is on and the supply

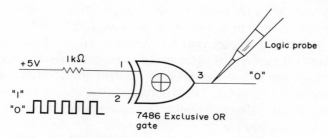

FIGURE 33   A SPARE EXCLUSIVE OR GATE USED AS AN INVERTER

connection to the device is unbroken. Although the probe clearly does not give a value to the indicated state, it does indicate that the supply connection is made; the actual value of the supply voltage has to be checked using a DVM. Similarly, touching the tip to pin 7 or 8 should cause the indicator to go out. If in either case the indicator remains dim, then a disconnection will be inferred and the printed circuit board tracks probed from the device pins back to the supply rails.

The logic probe is useful for verifying that a logic gate is functioning correctly.

Where circuit board space is limited, rather than try and squeeze extra gates onto the board use is often made of spare gates in devices used in a system. Take as an example the Exclusive OR gate shown in Figure 33. A 7486 contains four such gates in a single 14-pin package, of which perhaps only three are used. An inverter is needed in the system and the otherwise spare fourth gate in the package is utilised as one. Connecting a logic probe to pin 1 will indicate a logic "1" state as a bright display, while touching the tip to pin 2 will show pulse activity. The function of the gate should be such that its output is the input pulse train inverted, so that connecting the probe to the output should also indicate pulse activity. The fact that the pulse train is inverted in sense to the input cannot be determined from the probe display. If instead of pulse activity at the gate output, the display stays off, to indicate a continuous logic "0" state, then a fault exists which may either be due to a fault in the gate itself or to a short to ground external to the gate. The short may be a solder splash creating an electrical connection between the line connected to the gate output and ground, or a short to ground on an input pin inside any devices to which the output connects. To determine the actual fault, either the gate output pin can be isolated as outlined in Section 3.5 or alternatively, a current tracer can be used to discover which device on the line is sinking current.

In a microcomputer system a logic probe is a useful item of test equipment to check initially for static logic levels and to verify bus activity. Control bus lines may be tested to ensure that a fault on a critical control line is not preventing the system from operating, as will be the case if say the $\overline{\text{HOLD}}$ control input used on several microprocessors is being held permanently low. Continuity in a limited sense may also be checked using a probe; if, say, a memory device does not appear to be being selected, then its $\overline{\text{C E}}$ (chip enable) input may be tested for pulse activity with a probe and traced back along the printed circuit track to the address decoder ouput driving it. On circuit boards with a high density of devices, the track widths used are narrow and a hairline crack may develop on a line. By running a probe along a track, it is often possible to locate the break which cannot be seen with the naked eye.

5.2 LOGIC PULSERS

A logic probe can only be used to monitor activity at one single node in a network; to complement the logic probe, a logic pulser is used to stimulate a node by forcing it from one logic state to the other and back again. Logic pulsers are in-circuit stimulus devices which inject short duration, high current pulses into a node to force it to change state and then revert. Typically, a pulser will source or sink up to 0.75 ampere into a node for 300 nanoseconds; because of the short duration the IC being pulsed will not be damaged. The output drive stage of a pulser is tri-stated, so that under normal conditions touching the tip to a logic node in a network does not affect the activity. A push button, located on the pulser body, is manually pressed to inject a single pulse into a node under test. Located in the tip of a pulser is an indicator which is pulsed in sympathy with the output signal. Pulse stretching is used to enable the display indicator to have sufficient time to actually turn on and off and be viewed by the user because of the very short duration of the actual output signal.

A number of commercially available pulsers, such as Hewlett Packard model 546A, may be programmed to provide a sequence of pulses. This particular instrument can be set into any one of six different output modes using a single double-action switch. The switch can be pushed to provide a single output pulse and also pushed forward to lock it into a mode. By implementing a series of manual pushes of the switch and then pushing it forward to lock it, a preselected mode can be latched into the control logic. The model 546A can provide any one of the following:

(a) 1 pulse.

(b) 100 Hz continuous pulses.

(c) A burst of 100 pulses.

(d) 10 Hz continuous pulses.

(e) A burst of 10 pulses.

(f) 1 Hz continuous pulses.

The rates at which the display is pulsed are reduced to 10 Hz for the 100 Hz modes and 5 Hz for the 10 Hz modes.

The control logic in the 546A remembers the number of button pushes used to select a mode. Say, for example, that a burst of 100 pulses is needed, then this will involve pushing the select button twice and then locking it forward. Each push of the button generates a single output pulse; this information is stored in the control logic so that only a further 98 pulses are generated for the first burst, making a total of 100 pulses in all. Each 100-pulse burst takes 1 second, and the pulser then waits for another second before sending another 100-pulse burst. Releasing the button from its locked position will terminate its output mode, unless the pulser is in a burst mode, when it will complete its current sequence of pulses before unprogramming itself and tri-stating its output pin.

The output impedances of logic devices are sufficiently high to allow the pulser to force them to change state momentarily.

A logic pulser cannot, for example, be used to pulse a power supply line, because its impedance is far too low.

5.2.1 Using a logic pulser

A logic pulser is used to inject either single pulses or sequences of pulses into a logic system without the necessity to disconnect devices from the circuit.

Figure 34 illustrates a typical circuit used in microcomputers to give an audible signal. The driving gate may be either a TTL totem-pole device or an open collector output gate. TTL devices can sink more current than they can source, so to achieve a higher output power level from the loudspeaker or piezo-electric buzzer one end of it is

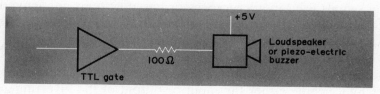

FIGURE 34   A "BELL" CIRCUIT OFTEN USED IN MICROCOMPUTER SYSTEMS

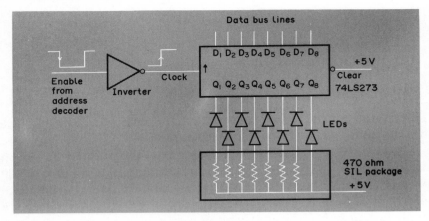

FIGURE 35   DISPLAY OUTPUT PORT USING A 74LS273 OCTAL D TYPE LATCH

connected to the 5-volt rail and the other end grounded via a 100-ohm resistor through the gate output. When the gate output is in the logic "1" state, no current flows in the circuit. By pulsing the input to the gate at a suitable rate, an audible tone will emanate from the loudspeaker. Assume that due to a fault no sound is emitted from the loudspeaker when the circuit is pulsed by the computer system. The driving gate and the speaker may be tested by connecting the logic pulser to the input of the gate and selecting either the 100-Hz burst mode or 100-Hz continuous mode. If all is well, the speaker will emit an audible tone at 100 Hz. If the test fails, then the pulser can be transferred to the output of the gate and the test repeated. If the test still fails, the pulser may be connected directly to the speaker and provided that its impedance is not so low as to prevent the pulser from operating, this should cause an audible tone to be heard. The pulser can thus be used to test every component in this simple circuit.

The display output port shown in Figure 35 is commonly used in microcontrollers as a simple means of displaying status information such as the part of a cycle that a sequencer may be in at any particular time. Data on the data bus is latched into the 8 D type latches when the clock input is taken to the logic "1" state. Conventionally, device enable signals used in microprocessor based systems are active low and to obtain the correct sense of signal the signal from the address decoder is inverted before being applied to the clock input of the 74LS273 device.

The operation of the output port may be verified using a logic pulser by touching either the input or output of the inverting gate. This will force an active clock edge onto the D type latches and whatever data happens to be on the data bus at the time will set the

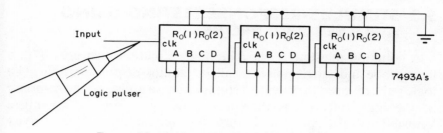

FIGURE 36   A 12-BIT ASYNCHRONOUS COUNTER

states of the LED indicators. If, for example, the data bus line connected to input D_1 happens to be a logic "0" at the time of pulsing, then the LED connected to output Q_1 will light up. Data bus activity normally changes so rapidly that pulsing the clock input several times will cause all the LED indicators to change state from off to on. If an LED remains off every time that the port is clocked, then it indicates one of several possible faults, such as a faulty data bus line, a fault within the 74LS273, or a faulty LED or resistor. Using the pulser directly on the output pin of the 74LS273 will have little effect, because the corresponding LED cannot turn on and be seen for the duration of the pulser output signal. Connecting a logic probe to the suspect output while the port is clocked, however, overcomes this problem, because of the pulse stretching logic built into it. If the output port pins do not have display indicators connected to them, a logic probe can be used in their place to verify each output in turn by clocking the port several times and noting changes of state on each output.

The Hewlett Packard model 546A can generate accurate bursts of 100, 10 or single pulses, which can prove useful when a precise number of clock pulses are needed to preset a counter, shift register or other sequential circuit.

Let us assume that we have the 12-bit asynchronous counter, using 7493A TTL 4-bit counters shown in Figure 36. We wish to preset the counter to give the output bit pattern 1000 1000 1000, which is 2184 in decimal. The 546A pulser is touched onto the input to the counter and programmed to generate bursts of 100 pulses. The operator counts 21 such bursts and removes the pulser from the circuit. The pulser is then reconnected and programmed for 10-pulse bursts and the operator counts 8 bursts before removing the instrument. Finally, the pulser is reconnected and the operator manually enters 4 pulses by pressing the pulse putton 4 times. 2184 pulses have been entered into the counter to set the output bit pattern to the desired value.

5.3 STIMULUS-RESPONSE TESTING USING A PULSER AND PROBE

A common method of checking logic circuits which uses both a logic probe and a logic pulser is called stimulus-response testing. The logic pulser is used to provide a stimulus into a node and the response of the logic network is then traced using a probe. Applying a knowledge of the circuit under test, a tester can trace signal activity along the propagation path using the logic probe.

Probably the simplest and most common use of stimulus-response testing is to determine continuity from the output of one gate to the input of another.

To check the circuit shown in Figure 37, initially the input to gate G_1 at the logic "0" level will be pulsed and the output of G_1 probed to ensure that a signal path exists through the gate. The pulser may be left on the input to G_1 or moved to its output and the probe touched to the input to gate G_2. If the line connecting the gates is open circuit, the probe will indicate a "bad" logic level and not respond to any line stimulus from the pulser.

It should be borne in mind that the actions initiated by the pulser will propagate along a signal path if continuity exists, and this in itself can give rise to either unexpected or unwanted system responses. Take, for example, the circuit given in Figure 37. Probing the suspect line for continuity will also cause the J-K flip/flop clocked from the output of gate G_2 to change state if the suspected line is not open circuit. The effect of changing the state of the flip/flop may trigger actions further along the signal propagation path and in turn may cause the inputs to G_1 to change and alter the state of the line being probed. This can be misleading if, when initially probed, the line indicates a logic "1" state on the probe and after pulsing suddenly changes to a logic "0" state. To interpret such a change, the tester must be aware of the complete circuit and its possible actions if he is to deduce meaningful results from the circuit responses.

Although stimulus-response testing of the type described is in widespread use for testing logic networks, the responses are easier to

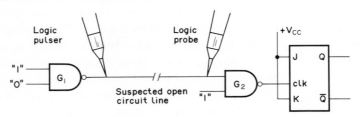

FIGURE 37   TESTING A PRINTED CIRCUIT TRACK FOR CONTINUITY

interpret in combinatorial networks than sequential circuits because of the time invariance of the former type of system.

5.4 THE CURRENT TRACER

A current tracer detects current flow along a printed circuit track by means of an inductive pick-up coil, located at its tip. Michael Faraday discovered the law of electromagnetic induction on August 29, 1831 and quite independently, a year later, the American physicist Joseph Henry discovered the same effect. The law as formulated by Henry states that an electromotive force (emf) is induced in an inductive circuit when the current flowing in the circuit changes. The equation that relates the induced emf to the current change is:

$$E = -L\,\frac{di}{dt}$$

where E is the induced emf, L is the inductance of the circuit and di/dt is the rate of change of the current. The negative sign is an indication of the law due to Heinrich Lenz, which simply stated says that the induced emf is in such a direction as to oppose the change of current causing it. Faraday demonstrated that if two circuits were coupled together by a magnetic path, then changing the current in one circuit caused an emf to be induced in the second. The changing current in the primary circuit caused a magnetic flux in the magnetic path which in turn caused an emf to be induced in the second circuit.

Faraday's law states that if an emf is suddenly applied to an inductive circuit, wound on an iron ring, then the emf can be related to the change in magnetic flux in the iron ring by:

$$e_1 = -N_1\,\frac{d\Phi}{dt}$$

where N_1 is the number of turns of the primary coil and Φ is the

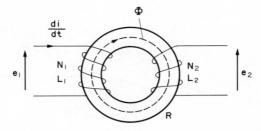

FIGURE 38   ELECTROMAGNETIC INDUCTION

magnetic flux. Combining the formula due to Henry with that of Faraday gives:

$$e_1 = -N_1 \frac{d\Phi}{dt} = -L_1 \frac{di}{dt}$$

from which

$$\frac{d\Phi}{dt} = \frac{L_1}{N_1} \frac{di}{dt}.$$

The same magnetic flux change causes an emf to be induced in the secondary winding, given by:

$$e_2 = -N_2 \frac{d\Phi}{dt}.$$

The secondary induced emf is hence

$$e_2 = -\frac{N_2}{N_1} L_1 \frac{di}{dt}.$$

If the primary winding is part of a printed circuit track, then in effect it represents a single turn winding, and N_1 equals one. Thus the emf induced in the secondary winding is given by:

$$e_2 = -N_2 L_1 \frac{di}{dt}.$$

N_2 is a constant and the induced emf e_2 is directly proportional to the rate of change of current flowing in the printed circuit track.

A current flowing along a printed circuit track produces a magnetic field and when the magnitude of the current changes, the magnetic flux changes. This change in flux is used in a current tracer to cause an emf to be induced in a search coil mounted in the tip of the tracer. The design of the search coil is critical, because it has to be constructed as small as is practical, so that when tracing a changing current in one circuit track it is not severely influenced by changing currents in adjacent tracks.

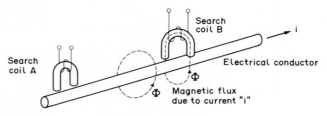

FIGURE 39   THE MAGNETIC FLUX PATTERN AROUND A CIRCULAR CONDUCTOR

A typical search coil consists of a horseshoe magnetic circuit, constructed from ferrite, with a coil wound onto it. A circuit board track can be considered as similar to an electrical conductor which causes a circular magnetic flux to be set up around it when a current flows. Search coil B is oriented such that the magnetic flux due to the current couples into its magnetic circuit. If current i changes, then the flux changes in sympathy, which causes an emf to be induced in the winding on search coil B. Search coil A, however, is oriented such that there is no coupling between the flux due to the current i and its magnetic circuit, and when the current changes there will be no induced emf in search coil A. The search coil mounted in the tip of a current tracer therefore has to be oriented to correspond with the direction illustrated for B, or little or no emf will be induced and the instrument will not detect changes in current. Usually a mark is placed on the tip which has to be aligned with the circuit board track being monitored to ensure maximum induced emf.

A current tracer can only detect changes in current and will give no indication at all if a constant current flows along a conductor. Digital circuits are pulse circuits in which signals are constantly changing from one logic state to the other, which in turn causes the current flowing in the electrical connections to change. The actual magnitudes of the currents may be small, but the time in which the change takes place is normally very short, giving rise to measurable values of di/dt. A current tracer such as Hewlett Packard's model 547A is sensitive to changes in current from 1 milliampere up to 1 ampere and has an adjustable sensitivity control mounted on it so that the intensity of the display lamp used to indicate current activity may be adjusted. In use, the control is adjusted by the user so that he can gain information about the relative magnitudes of currents flowing in different tracks from the intensity of the lamp. The instrument requires more practice, if it is to be used successfully, than say the logic probe or pulser, because the user has to gain experience in setting the sensitivity control and get used to orienting the tracer tip correctly in line with the track he is tracing. The high packing density of devices and tracks on modern printed circuit boards often makes it difficult to place the tracer tip on one track alone, and the indicator may display information not only about the track supposedly being monitored but also about an adjacent track.

The troubleshooting attributes of a current tracer are less obvious than those of the logic probe and logic pulser because it is used to trace current flow rather than voltage levels with which people tend to be more familiar.

5.4.1 Using the current tracer

For correct current tracer operation the tip must be perpendicular to the track being monitored and aligned in its direction. If a track is being monitored with the tracer and it bends at, say, right angles to itself, then the tracer must also be rotated to remain in line.

The tip of a current tracer is electrically insulated so that it can be placed directly onto a circuit. If the tracer is placed on one of the address bus lines of a working computer system, then the sensitivity control is adjusted until the display lamp shows at about half its maximum intensity. If the tracer is rotated in contact with the same location on the address bus line, the lamp intensity will be seen to vary to off and back to half intensity, provided the tool is correctly aligned initially. If the tracer is placed on a power supply rail, the intensity will be seen to increase due to the larger currents. In practice, when tracing current flow in other than power supply rails, it is easiest to set the sensitivity control to produce a dim lamp indication, which allows the instrument to detect small differences in current. To provide a useful indication, a current tracer incorporates pulse stretching logic so that the frequent and rapid change in current levels can be seen by the operator.

5.4.2 Stimulus-response testing using the logic pulser and current tracer

The current tracer is generally inferior to the logic probe for detecting circuit activity, because it is more difficult to use. However, for certain faults it has no equivalent and the short time required to become proficient in its use may be more than amply rewarded. In many situations, such as an internal short to ground inside an IC, the current level flowing into the short may be static and hence cannot be detected by the current tracer alone. To overcome this problem, the tracer is often used in conjunction with a logic pulser which stimulates the faulty line and the current change due to the pulser output can then be detected.

When using a current tracer in conjunction with a pulser to fault find on a data line it is useful to set it from the pulser output as shown in Figure 40. The sensitivity control on the tracer is adjusted to give a relatively dim indication and at roughly midway in its rotation, which means that the injected current is probably between 10 an 100 milliamperes. This is the sort of current level to expect from a TTL or CMOS output gate driving a line to which are connected several gate inputs in the logic "0" state. If, for example, there is a hard short to

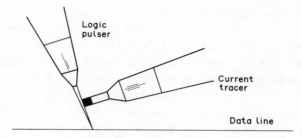

FIGURE 40   SETTING UP A CURRENT TRACER WHEN USED WITH A PULSER

ground on the data line, the pulser output will inject about 1 ampere of current and the tracer's sensitivity control will give a bright indication at any setting. Experience with a current tracer allows educated guesses to be made about the probable current level and its source.

5.4.2.1 WIRE-AND NODES

One of the most difficult problems encountered in troubleshooting digital ICs is a stuck wire-AND node.

The gates G_1, G_2 and G_3 are all open-collector output NAND gates whose outputs are coupled together and pulled up to V_{cc} by a pull-up resistor. The configuration is referred to as a wire-AND because the Boolean expression for H in terms of the inputs A to F is:

$$H = (\overline{A \cdot B}) \cdot (\overline{C \cdot D}) \cdot (\overline{E \cdot F}).$$

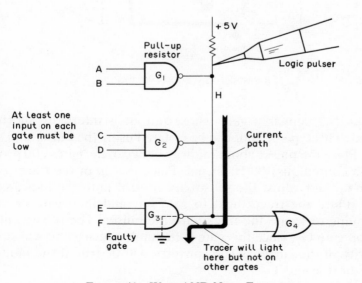

FIGURE 41   WIRE-AND NODE FAULT

If one of the output transistors in the gates G_1 to G_3 becomes permanently shorted to ground, then a logic probe can be used to ascertain the fact that the node is always low, no matter what the state of the inputs A to F, but it cannot be used to decide where the fault lies. The fault can be in any of the driving gates G_1 to G_3, or be a short in gate G_4, or be due to a short to ground on the line itself.

Let us assume that gate G_3 output is causing the short, then the node can be driven by a logic pulser and the current path to ground traced using a current tracer to pinpoint gate G_3 as the faulty device. The current tracer in this situation provides a rapid means of determining which device of several is causing the problem; the alternative to using a current tracer here will be to isolate the output of each driving gate in turn to see if the fault clears, which is clearly a time-consuming task.

5.4.2.2 GATE-TO-GATE FAULTS

When a low impedance fault exists between two gates, the current tracer and logic pulser can be used to establish the faulty device quickly.

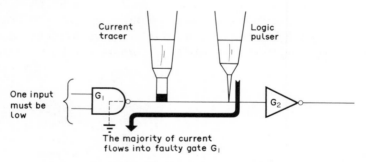

FIGURE 42   GATE-TO-GATE FAULTS

Figure 42 depicts a gate whose output is internally shorted to ground. This type of fault can be isolated using the pulser and tracer tools. Place the pulser about midway between the gates and program it for a continuous 100 Hz signal. Place the tip of the tracer on the pulser tip and adjust the sensitivity control until the indicator just lights. Place the tracer next to gate G_1, and then gate G_2, while continuing to excite the track with the pulser. The tracer will light only on gate G_1, since this gate is the faulty device and will sink the majority of the current. The converse will be true if the fault is an input short in gate G_2.

5.4.2.3 SOLDER BRIDGES AND CABLE PROBLEMS

When checking printed circuit tracks which have suspected short circuits due to solder bridges or shorts due to unetched tracks, the current tracer is started at one end of the line where a pulser is used to excite it continuously and then moved along the track to trace out the major current path. This type of fault often occurs during the production of circuit boards where a short may have been caused between two lines during a manual soldering operation or after passage through a reflow soldering machine. The same technique can sometimes be used when troubleshooting shorted cable assemblies.

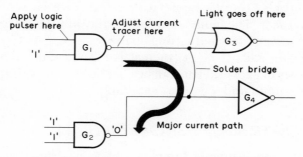

FIGURE 43   SOLDER BRIDGE ON A PRINTED CIRCUIT

The example in Figure 43 shows an incorrect path due to a solder splash across two tracks. As the tracer follows the track from the output of gate G_1, towards gate G_3, the indicator remains alight until it passes the bridge. The indicator then extinguishes, showing that the current has found some other path than the track from G_1 to G_3. A visual inspection of the region where the indicator goes out should reveal the fault.

5.4.2.4 MULTIPLE GATE INPUTS

A common type of logic structure is where one gate output drives several gate inputs as shown in Figure 44. The majority of logic gates have input protection diodes on all their inputs, either to protect the device from a build up of static charge, as in the case of buffered CMOS devices, or to speed up the gate response, as is the case where Schottky diodes are used in TTL devices. These diodes can become short circuit and cause a low impedance path to ground. Several gate inputs may be connected onto one line and the problem becomes one of determining which input is at fault.

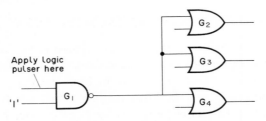

FIGURE 44   MULTIPLE GATE INPUTS

The logic pulser will be applied to an input of gate G_1 and the current tracer set up to just give an indication at gate G_1 output. Then each of the gates G_2 to G_4 inputs will be tested for current flow with the current tracer. The faulty input will be the only device to cause the tracer lamp to light. If when setting up the tracer at gate G_1 output the tracer indicator fails to light, then it is a good indication that the problem exists in gate G_1. This can be checked by using the pulser and tracer as described under gate-to-gate faults.

5.4.2.5 POWER SUPPLY SHORT CIRCUITS

One of the most difficult faults to isolate is a short circuit of the power supply rails on a printed circuit board. The 5-volt supply used by the majority of logic families is connected to almost every IC on the board, and is decoupled by ceramic and tantalum capacitors at regular intervals. On a circuit board containing say twenty ICs there may be as many as ten to fifteen decoupling capacitors across the supply and ground. It is unusual for an IC to provide a short circuit internally between the supply rail and ground without it being visually apparent. If an IC has been damaged to this extent, the centre of the DIL package will normally have erupted and can be readily seen. Although not unheard of, this type of fault is relatively rare. However, short-circuited decoupling capacitors are not an unusual occurrence, and because they are all paralleled across the supply the isolation of one defective capacitor is usually achieved by disconnecting each one in turn. This can be a time-consuming task on a densely packed circuit board.

This type of fault can be located by disconnecting the supply from the circuit board and using a logic pulser to excite the supply rail and tracing the current path with a current tracer. A separate power supply (the one disconnected from the board will suffice) has to be used to power the pulser and current tracer and the ground connection left connected to the circuit board. The pulser is then programmed into its 100-Hz continuous mode and the current tracer set up as

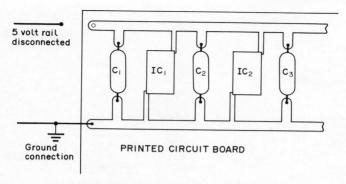

FIGURE 45   POWER SUPPLY SHORT CIRCUIT

illustrated in Figure 40. The tracer is then moved along the supply rail track until a point is reached where its indicator goes out. It may be that the supply rail splits at this point and the tracer then has to be relocated onto the track into which the current is flowing. Eventually the tracer should reach a point on a track which does not branch where its indicator extinguishes. The faulty component can then be isolated to this area and may be a decoupling capacitor or IC. If the IC is socketed, then the first step is to remove it and check again whether the fault clears. If the fault persists, then suspect a decoupling capacitor in the vicinity of the isolated area.

In Figure 45 assume that the decoupling capacitor C_2 has short-circuited. The pulser and current tracer will be powered from the 5-volt supply which has been disconnected from the circuit board. The pulser will be applied at the supply connection point to the circuit card and the tracer moved along the supply track until its indicator goes out. This will occur in the example just past capacitor C_2, indicating that it is the faulty component.

5.5 LOGIC COMPARATORS

A limited number of complete ICs can be tested in-circuit using an instrument called a logic comparator. The comparator is fitted with a card containing the same device as that to be tested, and a clip from the instrument is connected onto the device under test. The comparator then takes all the input signals from the device under test and applies them to the similar device plugged into it. The outputs from both devices are then compared in Exclusive OR gates and if any of them differ the resulting error signal is stretched and used to illuminate an LED.

The comparator removes the burden from the troubleshooter of analysing long, complex signals by automatically determining if the

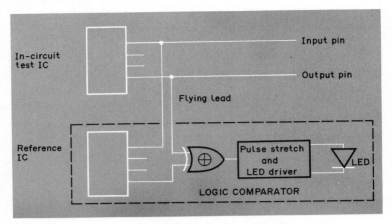

FIGURE 46   CONCEPTUAL LOGIC COMPARATOR

signal is a good "high" or "low" level and that the level is correct according to the truth table of the device being tested.

It must be borne in mind when using a logic comparator that it will load the circuit under test which, in a circuit where the maximum fan-out of a device is already in use, can cause erroneous behaviour. The Hewlett Packard 10529A comparator, for example, will apply an extra 1.2 input loads to any input it is connected to, but only 0.1 of a TTL load to any output. In most situations these additional loads will not affect the performance of the circuit under test, but where analogue components are used, such as in a monostable circuit, the additional loading can affect the circuit timing adversely.

Sequential circuits can also present problems to comparators because their outputs depend upon previous inputs, and the reference device in the comparator will not in general have had the same "prior" set of inputs. In such cases it is necessary to reset the device being tested before making any comparisons.

The wide range of devices in any logic family means that some devices are untestable by a logic comparator; usually these include any lamp drivers/decoders such as the 7447 TTL. Another disadvantage of the logic comparator is that a separate reference card has to be stocked for each type of device that may have to be tested. There are currently seven different TTL families which cover several hundred different devices, which can make stock holding of reference cards a formidable task.

5.6 THE LIMITATIONS OF HAND-HELD TOOLS

Logic probes, pulsers, current tracers and logic comparators have for a long time been the mainstay of digital troubleshooting. They are, however, restrictive in the sense that a pulser can only excite a limited number of nodes in a network simultaneously and a logic probe can only examine a single network node at any one time. The probe can give useful information about the static condition of a node or indicate that it is being subjected to pulsed activity; it cannot, however, give any meaning to the pulsed activity. Hand-held tools play an important role when analysing faults in conventional logic networks. They are, however, limited in scope when applied to bus structured systems where information appears on a large number of lines simultaneously and in a time sequential form. To comprehend the operations that occur in a microcomputer system, a tester has to have equipment which can capture and display, in a suitable form, the information on many lines, and have the ability to qualify the information he wishes to study. Clearly, simple hand-held tools cannot perform these functions, and this has led to the development of test equipment specifically designed to cope with the demands of troubleshooting in complex, bus structured systems.

Chapter 6

Logic Analysers

In a microcomputer system the information relevant to a single operation is contained over a short period of time on many lines. Every time a CPU reads the opcode portion of an instruction from memory, it has initially to place the address of the opcode word in memory on the address bus. It then issues a *Memory* request control bus signal (in the case of an I/O mapped I/O system) and sends a *Read* request control signal to the selected memory device. The enabled memory device then places the opcode word onto the data bus from which it is taken by the CPU and placed in its *Instruction Register*. Typically, this instruction fetch cycle has to be performed every time an opcode or operand is read from memory and it takes three system clock cycles to complete.

At the rising edge of clock pulse T_1, the contents of the program counter are put onto the address bus lines, which in a typical 8-bit microprocessor system will contain 16 lines. The states set onto these lines are each allowed half a clock period to stabilize before memory request and read controls signals are issued. These are indicated as $\overline{\text{MREQ}}$ and $\overline{\text{RD}}$ in Figure 47 to indicate the fact that, conventionally, they are active low signals. The mnemonics used to represent these signals differ between microprocessors. In the case of a memory mapped device, the memory request control line may not even exist. The microprocessor allows a further $1\frac{1}{2}$ clock periods before sampling the data bus lines on the rising edge of clock pulse T_3. This allows the memory device time to decode the address and place the contents of the specified location on the data bus. It is expected to have put its information on the data bus lines before this time and that it is stable. If a memory device cannot respond within this time period, it has to issue a $\overline{\text{WAIT}}$ signal to the processor to prevent the CPU reading bad data off the bus. Shortly after sampling the data on the bus, the CPU releases the $\overline{\text{MREQ}}$ and $\overline{\text{RD}}$ control signals and removes the address from the address bus. All microprocessors carry out memory read operations in a similar fashion to that illustrated by Figure 47,

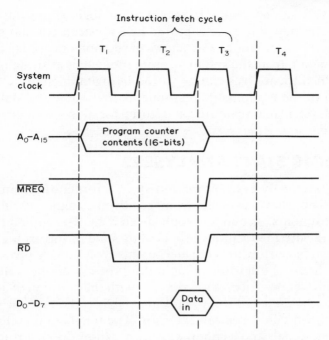

FIGURE 47   AN INSTRUCTION OPCODE FETCH CYCLE

although they may differ slightly in detail. For any particular device, the timing diagram for an instruction fetch cycle can be found in its data sheet.

A study of Figure 47 indicates that the total information needed to decode the referenced address, the type of operation being carried out, and the data relevant to that operation is only available in a stable form during the rising edge of clock pulse T_3. Thus if we can capture and store all the states of the address bus, control bus and data bus at this time, then we have all the information needed to interpret the operation being carried out. To actually achieve this, however, we have to store the states of 16 address lines, 8 data lines, at least 2 control lines, and synchronise the data capture to the system clock. We are thus faced with at least 27 lines of relevant information to capture and store. A further problem is that this timing may only apply to an opcode fetch, and when an operand is read from memory the address is left on the bus for a longer period and the data bus is sampled on the falling edge of the T_3 clock pulse. Fortunately, most microprocessors issue some form of control signal which informs any equipment connected to it that it is carrying out an opcode fetch, rather than reading an operand from memory.

The actual waveshapes of the signals present on the system busses are relatively unimportant. It is only their logic states during the period when their states are captured by an instrument that provide the necessary information. This allows the equipment used to capture the information from the system to store the signals as straight binary signals which themselves can be stored in digital memory elements contained in the equipment. Instruments that allow the states of a computer system to be captured and stored for subsequent display are called *Logic State Analysers*.

6.1 LOGIC STATE ANALYSERS

Logic state analysers capture and store information from digital systems. They have evolved from relatively simple, oscilloscope based instruments to complex, sophisticated analysis equipment.

To appreciate the facilities provided by state of the art logic state analysers, it is worthwhile considering the evolutionary process that led up to them. The starting point in this process was the well-established oscilloscope. An oscilloscope can with the addition of an input multiplexer be used to display many more signals than say the normal dual-channel oscilloscope can cater for. The timebase is common to all the input signals and its ramp waveform causes the electron beams to transverse across the faceplate from the left-hand side to the right. A similar ramp is used on the vertical Y axis and at certain points on it, one of the many input signals is sampled and displayed as a point on the screen. Up to say 8 input signals can be accommodated using this multiplexing technique which will allow 8 signals from a computer system to be displayed on the screen. The signals used to trigger the oscilloscope's timebase will be the computer system clock or a signal derived from it.

Information as presented on the oscilloscope in Figure 48 will be undecipherable because of the rate at which the displayed data changes, the fact that the information is presented as a set of squarewaves, and the trigger point is undefined. A single timebase

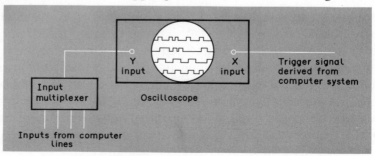

FIGURE 48   A MULTI-INPUT CHANNEL OSCILLOSCOPE

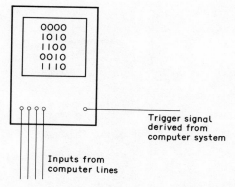

FIGURE 49   A MULTI-INPUT OSCILLOSCOPE WITH BINARY DISPLAY

scan of information can be taken and stored using a storage oscilloscope which removes the first problem. There still remains the
difficulties of deciding what the displayed data refers to in terms of the
computer operations and what event triggered the display. The
display can be made more readable by displaying the data in the form
of "1"s and "0"s rather than squarewave patterns, but this improve-
ment still leaves the essential problems of deciding what the displayed
data represents.

The display shown in Figure 49 is far more readable and represents
the hexadecimal characters 0, A, C, 2 and E. Address values and data
words in an 8-bit microcomputer system are usually given as hexadecimal characters. If more than 4 bits were to be displayed, such as
an address and a corresponding data value, the binary pattern would
be split into groups of 4 bits to improve readability (or groups of 3
characters if octal were in use). The following short list could represent a display of addresses and data values in an expanded form of the
system shown in Figure 49.

0000 0000 0000 0000	0011 1110	0000 3E
0000 0000 0000 0001	0000 0110	0001 06
0000 0000 0000 0010	1101 0011	0002 D3
0000 0000 0000 0011	0000 1000	0003 08

Each 16-bit address is shown as 4 groups of 4 bits, and the data is
shown as 2 groups of 4 bits. The equivalent hex codes are given on the
right-hand side.

The information no longer needs to be displayed as analogue
signals and the need for a conventional type of storage oscilloscope is
removed. In its place, RAM can be used inside the instrument for the
storage of addresses and data. The contents of the RAM can be

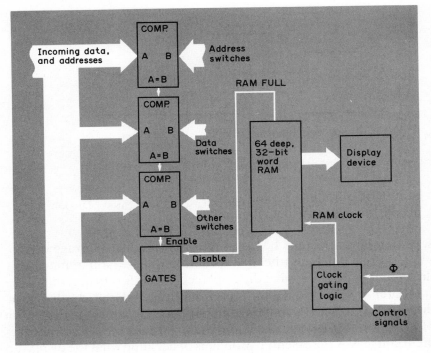

FIGURE 50 A LOGIC STATE ANALYSER

non-destructively read and displayed on a conventional type of oscilloscope for indefinite periods.

When testing a microcomputer, the troubleshooter is interested in viewing certain parts of a program and not just any circuit activity that happens to be captured by the analyser. He therefore requires some form of qualifying logic which will only capture information from the system under test for certain specific conditions; usually these will be the program sequence starting from some known address. The qualifying logic takes the form of a series of logic comparator circuits, into which are fed the conditions which must be satisfied before the analyser begins to capture information from the system under test. On the early logic state analysers, the qualifying information was fed into the analyser using a set of three position switches to represent the states "1", "0", and "don't care".

A block diagram of an analyser which meets these requirements is shown in Figure 50. All the address bus lines from the system being tested are connected into a 16-bit comparator, the other 16 inputs coming from a set of tri-state switches. Similarly, the data bus is fed into a similar comparator along with the states of 8 tri-state data switches. Usually provision is also allowed to input up to a further 8 signals, which couple into a third comparator.

When the address on the bus coincides with that set on the address switches, the A = B output from the top comparator sets. This then enables the second comparator connected to the data bus lines. If the data value on the data bus also agrees with the value set on the data switches, the output A = B of the second comparator sets which in turn enables the third comparator. If at the same time the states of the 8 other undefined lines from the system under test agree with the states set on the switches, its output A = B sets and enables the set of gates which allows the address, data and other signals through to the 32-bit wide RAM. By allowing "don't care" states to be set on the comparator input switches, information may be stored from the system under test on address values alone. If, say, all the data switches and switches connected to the lower comparator are set to "don't care", then when the incoming address agrees with the switch settings its A = B output signal will ripple through the other two comparators and enable the RAM input gates. In a similar fashion, data can be captured using only a data value as the qualifying trigger.

The RAM shown can store up to 64 complete pieces of information, each of which consists of a 16-bit address, an 8-bit data value and the states of another 8 lines from the system. This requires that the RAM be able to store words 32 bits wide. When the RAM has been filled with 64 system transactions, it generates a RAM FULL signal which disables the input gates. All 64 words in the RAM can then be shown on a display device for analysis.

Frequently the operator only wishes to view certain bus transactions such as I/O operations. To achieve this, the RAM clock is derived from the clock of the system under test but qualified by additional control signals. As an example, the RAM may only be clocked when an I/O request signal occurs, so that only I/O operations will be stored in the RAM. The user can thus be selective about the type of information captured from the system by being able to qualify not only address and data bus information, but also the types of transactions that are to be stored.

The system as illustrated in Figure 50 still, however, has one drawback; it can only store and display events which occur after the trigger word set on the comparator switches. Thus only events delayed in time can be captured from the system under test. When a system goes wrong, the solution can often be located if the events that led up to the erroneous event can be captured and subsequently displayed for analysis. This is called pretriggering and permits an

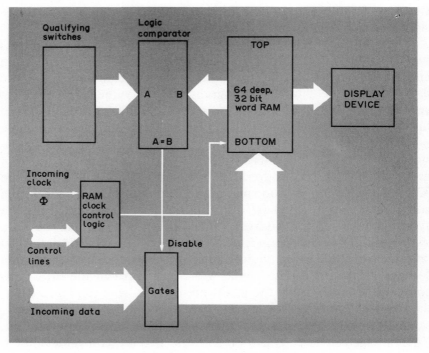

FIGURE 51   A LOGIC STATE ANALYSER WITH PRETRIGGER CAPABILITY

operator to capture and view events prior to the trigger word set on the qualifying switches as well as delayed words. The diagram in Figure 50 can be rearranged to allow for pretriggering.

In the rearranged form of the analyser, one set of information fed into the logic comparators comes, as before, from a set of qualifying switches, but the other set of information is now taken from a RAM location rather than from the system under test busses directly. The same form of clock qualifying logic as used in, Figure 50 is still retained to clock information into the RAM, which behaves as a *First In-First Out* (FIFO) register. The essential difference is that information from the system being tested is constantly being clocked through the RAM until the trigger word set on the qualifying switches coincides with the same data pattern in the RAM location coupled into the logic comparators. When this occurs, the input data gates are disabled and the contents frozen. The RAM data may be directed onto a display device for analysis by the operator.

The RAM location used for a logic comparison may be selected to be any one of the 64 possible words available. Say, for example, that the top location in the RAM is used for the comparison. When the bit pattern clocking up through the RAM into this location agrees with the set switch pattern, the input gates will be disabled. From the top downwards, the data stored will correspond to events that occur after the trigger word (all delay), and the system will function in an analogous manner to the scheme described in Figure 50.

If now the RAM location used for comparison is the lowest location at the bottom, then when its contents agree with the qualifying switch settings the RAM will be frozen and all the information stored in it will represent events that preceded the trigger word (all pretrigger events).

By selecting a RAM location between the top and bottom for comparison, when the trigger word is detected the RAM will freeze and contain the trigger word itself, several delayed values and several pretrigger values. The scheme can clearly satisfy the need to capture and display events that precede some trigger event and as depicted could allow up to 63 pretrigger computer transactions to be displayed. Where one needs to go back beyond this limit, the event that occurred 63 transactions before the trigger event of interest can now be set on the qualifying switches and a further 63 events prior to that captured and displayed. Simply by noting the intitial contents of the displayed information, the user can step back from an event as far back as necessary in blocks of 64 words.

The systems represented by Figures 50 and 51 are conceptual descriptions of logic state analysers, which for the sake of clarity do not indicate the way that the logic comparator handles "don't care" states, or the way in which the RAM information is sent to the display device for viewing by the user. In practice, most analysers use logic devices to handle incoming data to provide the speed necessary to cope with high data rates, and a microprocessor to provide overall control.

The incorporation of a microcomputer as the overall control element in a logic state analyser allows a number of desirable features to be included in the instrument. Figure 52 depicts in block diagram form the major components that are typically used to construct an analyser, which retains high-speed logic elements for data capture from the *System Under Test* (SUT), and a microcomputer to exercise overall supervisory control.

The information taken from the SUT's busses is buffered to minimise the loading of the instrument on the equipment being monitored. The outputs from the buffers are fed into a set of logic comparators to detect a preset trigger pattern and into a FIFO RAM. The sets of qualifying switches used on earlier analysers are now replaced by output ports from the microcomputer. The trigger word address, data and uncommitted inputs bit patterns are entered into

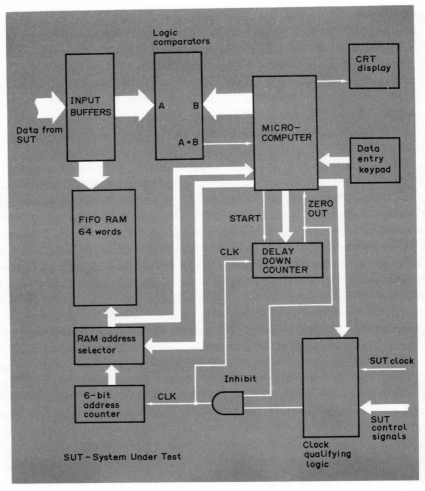

FIGURE 52 A MICROCOMPUTER BASED LOGIC STATE ANALYSER

the microcomputer from a keypad along with the number of pre-trigger events that are required before the trigger word. The FIFO RAM can only contain 64 words of information, and from the keyed-in pretrigger number the microcomputer can be programmed to calculate the number of delayed events that have to occur to fill the RAM. If, for example, a pretrigger value of 10 is dialled into the system, the computer will calculate that there will have to be 53 delayed events stored in the RAM to fill it up. This value is loaded into a delay down-counter from an output port of the microcomputer.

The type of system transactions that are to be stored are also fed into the computer from the keypad, and the information is used to set up the clock qualifying logic that controls the FIFO RAM addressing.

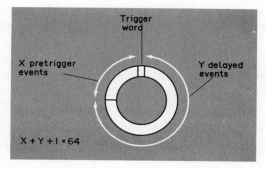

FIGURE 53   REPRESENTATION OF EVENTS IN THE CIRCULAR RAM

Having keyed in all the necessary information, the analyser is commanded to start, again by a keypad entry. The data selector which determines the source of the FIFO RAM addresses is set up to connect the 6-bit address counter to the RAM.

When the information from the SUT agrees with the qualifying words sent to the logic comparators from the computer, a signal from the comparators flags the computer that the preset trigger event has been detected. The computer then reads in from the 6-bit RAM the address at which the trigger event has been stored and starts the delay down-counter. After the calculated number of delayed events have passed, a ZERO OUT signal from the down-counter inhibits any further event storage by preventing the 6-bit address counter from being clocked. This signal is also read by the computer to indicate that all the required events are now stored in the RAM.

The RAM acts in effect as a circular memory store, and when data has been written into its lowest location the address counter resets and begins storing the next item of information in the highest location. To keep track of events in the RAM, the computer makes use of the address read from the RAM address selector when the logic comparators indicated that the trigger word has been recognised.

The total number of storage locations within the RAM is known (in our example, 64). From a knowledge of the address of the trigger word in RAM, and from the number of keyed in pretrigger events required, the computer has all the information it needs about the location in the RAM of all the pretrigger, trigger and delayed events.

For high-speed operation, the output from the logic comparators will be used to latch the RAM address which corresponds to the stored trigger word, rather than incur a time delay penalty of having the microcomputer read in the RAM address.

Having now frozen the data in the RAM, the microcomputer can access it at a slower rate by taking control of the RAM address selector, and setting the RAM address from one of its output ports. The analyser will now revert to a display mode of operation where the keypad can be used to control the display of stored data, typically on a CRT display.

The inclusion of a microcomputer in the system now permits a wide range of display formats. The binary patterns previously used on analysers can be upgraded, so that actual hexadecimal numbers are displayed for addresses and data values. The system becomes a little more "user friendly" by displaying the stored information in a more readable form.

For a specific microprocessor, the computer can be programmed to "disassemble" the hexadecimal codes and present, on the CRT screen, the mnemonic codes in place of their hexadecimal numbers. This drastically improves the readability of a displayed program segment and makes the task of debugging a program easier and faster.

The combination of keypad data entry, microcomputer and CRT display has permitted the development of ergonomically designed logic state analysers. The CRT display can be used to prompt a user, when setting up the parameters for data capture, by displaying on the screen a table of labelled items of information needed to configure the comparators and clock qualifying logic, and to inform the user if he enters an invalid parameter. In the display mode, usually the screen cannot handle all the stored information at the same time, and scroll keys are used on the keypad to scan through the data. The trigger word is frequently highlighted in reverse video and the scrolling operation is restricted to prevent overrunning past the last delayed event into the first pretrigger event. The operator can thus only view information as it occurred in the system and will not be confused by a display that suddenly jumps from an event that occurred after the trigger word to one that occurred before it.

Part of a typical disassembled display on a logic state analyser programmed to analyse Zilog's Z80 microprocessor could be:

ADRS	OPCODE/DATA	EXTERNAL
0D00	LD A,06	1111 0000
0D02	OUT(08)	1111 0000
0808	06 OUTPUT	1111 0001
0D04	CALL 0F06	1111 0011
0FFF	0D WRITE	1111 0011
0FFE	07 WRITE	1111 0011
0F06	IN A,(02)	1111 0001
0202	4C INPUT	1111 0000
0F08	LD (HL),A	1111 0000
2000	4C WRITE	1111 0000

At first sight, even to someone familiar with the Zilog assembly codes, this listing may appear confusing. The analyser not only displays the programmed instructions, but also the data and address bus transactions that accompany the instructions. To appreciate the displayed listing, the user has to have some degree of familiarity with the manner in which the particular microprocessor, in this case the Z80, operates.

The analyser was triggered using address 0D00 as the trigger word, and leaving both the data bus information and the states of the 8 external inputs as "don't care" states. The data stored from memory locations 0D00 and 0D01 was 3E in 0D00 and 06 in 0D01. The controlling microcomputer has been programmed to disassemble Z80 codes. It presents on the display under the heading OPCODE/ DATA the mnemonics LD A,06 which means load the A register in the CPU with the value 06. On the second line of the disassembled program this data value is sent to output port 08 with the OUT(08) instruction. When an I/O transaction occurs in a Z80 based system, the I/O port address is sent out on only the lower 8 address bus lines, but it is also duplicated on the upper 8. Thus the next line displayed on the analyser shows the port address (doubled up) as 0808 appearing on the address bus and the actual data value sent to that port as 06 on the data bus lines. To illustrate that the data bus value is being written to an output port, the data value (06 in this case) is followed by the word OUTPUT.

In the fourth line of the display at address 0D04, a CALL is made to a subroutine at location 0F06. When the program was written in mnemonic form and then assembled into machine code, this called address was probably given a symbolic label. The analyser cannot disassemble this information, because it is lost in the original assembly operation; the best the analyser can achieve is to give the address of the first instruction in the subroutine. A CALL instruction takes 3 bytes of memory space to store the opcode and the 2-byte address of the subroutine which leaves the next instruction in the main program at location 0D07. This address has to be returned to at the end of the subroutine and must be stored on the *Stack*. It has been assumed in this example that the *Stack Pointer* has been initialised to location 1000 in RAM. The return address (0D07) is pushed onto the stack by first decrementing the stack pointer to point to location 0FFF, and then the upper byte of the program counter (0D) is loaded onto the stack. The stack pointer is again decremented to point to location 0FFE and the lower byte of the program counter (07) loaded onto the stack. These two data bus transfers are shown on lines 5 and 6 of the analyser display. The program counter is overwritten with the start address of the subroutine at location 0F06 which is shown in line 7 as containing an INPUT instruction from I/O port 02. The data read from this port, along with the doubled up I/O address, is illustrated in line 8 as an apparent address of 0202 and a data value of 4C hexadecimal, read into the CPU's A register. The operation is qualified on the display as an input read by following the actual data value of 4C with the word INPUT.

The penultimate line of the display illustrates the value read into the A register being loaded into a memory location pointed to by the HL register pair. It has been assumed that the HL register pair had been previously loaded with the hexadecimal address 2000, so that the final line in the display shows the value of 4C being written to location 2000.

A possible argument against the form of display illustrated is that it tends to be cluttered with data bus transfers, which do not appear in an assembly listing of the program that is being debugged. The user has to exercise reason when deciphering he display an cannot just skip through a program listing looking for possible differences due to programming errors. The rebuttal to this argument is that the analyser displays important information about the true system bus transfers that may play a major role in detecting why a system is going wrong. The display, for example, gives the addresses of the stack that are being used which for any number of reasons, may themselves be erroneous. Memory may not have been located at the supposed stack locations, or the stack pointer may have been pointed elsewhere. The actual locations in use may have been used to store a program, or have been used as a data storage area. The actual data sent to I/O ports is also displayed and these can provide vital clues as to why a system may be performing incorrectly. In the example, when input port 02 was read, the value read into the CPU was 4C. This may not correspond to the value that it was thought the system was reading. It can be tested at the external inputs to port 02 to confirm or deny its accuracy. The data being fed into the port may be wrong, due to some external fault, or the port itself may be faulty. If test switches are connected to an input port and a display device connected to output ports, the analyser provides a means of testing the I/O connections.

Logic state analysers developed from plug-in oscilloscope modules which appeared in 1973 to the self-contained type of instrument that could perform disassembly on a specific microprocessor, which became available during 1977. During this four-year span, the capabilities of logic state analysers improved dramatically from instruments that were essentially difficult to set up and interpret, to equipment that was both easy to use and understand.

Logic state analysers are primarily used to assist in software debugging by tracing and displaying the state flow in an algorithmic state machine such as a microcomputer. All operations in an algorithmic state machine occur in synchronism with a system clock, which is used to capture and, along with other control signals which are also synchronous with the system clock, to qualify the type of information stored and displayed.

6.2 LOGIC TIMING ANALYSERS

Events sometimes occur in a system asynchronously to the system clock and they can cause erroneous behaviour; these events are often referred to as "glitches", and can arise from internal timing errors or from sources outside the computer system. Logic state analysers only capture information from a system in synchronism with the SUT clock and cannot detect the passage of a glitch. A state analyser samples the system busses only during edge changes of the clock. Any change in state of any bus line that occurs at any time other than during clock edges is not captured and stored. To permit the timing of a system to be investigated, and glitches sought for and captured, analysers have been developed which use their own internal clock as a timing and sampling source, which runs asynchronously to the SUT clock. This type is known as a *Logic Timing Analyser* and is primarily used to display timing diagrams of a system. The need for logic timing analysers grew from the basic limitations of an oscilloscope to trace and record waveforms simultaneously from many bus lines. Currently logic timing analysers will capture and display up to 16 timing waveforms using internal clock rates up to 200 Mhz. It must be stressed that the information stored and subsequently displayed is NOT an actual analogue representation of the waveforms being examined. The information is stored in internal RAM and displayed as idealised squarewaves. The RAM may be typically 1024 or 2048 bits long for each input, and the sampling rate may be selected from say a bottom limit of 2 Hz up to the maximum capability of the analyser. There is clearly a trade-off, because of the finite size of the storage RAM, between the sampling rate set into the instrument and the amount of SUT activity finally displayed. If we were analysing a system with a 500-nanosecond clock and set a sampling time of 100 nanoseconds on the logic timing analyser which has a storage RAM containing 1024 locations, then only 204 system events concurrent with the SUT clock can be captured and displayed. If the sampling time is decreased to 25 nanoseconds, then only 51 SUT events can be captured.

There are two types of asynchronous timing measurements that are applied to logic systems: the first is *Functional Timing Analysis* and the second *Parametric Timing Analysis*. If we are concerned that event A takes place during event B, then we are applying functional timing analysis, which simply shows the correlation in time between the events. If our primary concern is that event A takes a particular time to occur, or event B occurs at some definite time in relation to event A, then we are performing parametric timing analysis where the precise timing of events is of importance. The internal clock in a logic timing analyser used for parametric timing measurements has to run significantly faster than one used for functional timing measurements. This speed requirement, and the necessity to equip the analyser with logic devices capable of handling high data rates, significantly increases the cost of a parametric timing analyser over a functional timing analyser.

Glitches or spikes may occur due to "race conditions" in logic circuits, through poor design of part of a system, or be induced due to noise or from an external source. A glitch is usually defined in a logic timing analyser as a pulse the width of which exceeds 5 nanoseconds but is less than the sampling period set on the instrument. Thus if the sampling time is set at 50 nanoseconds, then any pulses on any of the analysers input lines greater than 5 nanoseconds but less than 50 nanoseconds in width will be treated as a glitch. Separate capture and storage is used in a logic timing analyser for glitches and the normal system events being recorded. When displayed, the two sources of timing information are merged and glitches are displayed as bright up pulses on the timing waveforms. On functional timing analysis machines, glitches are displayed as bright pulses, the width of which is preset by the sampling time set into the instrument. Thus if two glitches occurred, one of 30 nanoseconds width and the other of 70 nanoseconds width, and the sampling time has been set at 100 nanoseconds, then on the eventual display both glitches would appear of equal duration.

In practice, logic timing analysers are used to view control bus signals and I/O activity which are the predominant areas in which timing problems occur. Many companies manufacture devices which ostensibly are microprocessor compatible, but which have been designed to interface directly into relatively slow microcomputer systems. If one of these devices has to be interfaced into a faster microprocessor system, then attention has to be paid to the relative timing requirements of both. Control pulse lengths may have to be stretched to ensure that the interfaced device is operated within its guaranteed limits, and any clock signals used by the device may have to be derived from the computer system clock by frequency division logic. Relative timing skew can occur between the device and the system due to propagation delays through the division logic and due to the set-up times of the device itself. Particularly during the development phase, the design may appear correct but the system will misbehave due to an overlooked relative timing skew. A logic timing analyser can then be used to analyse the timing signals for any glitches due to a race condition, or to check the set-up times actually applied to the device.

An example of glitch detection can be illustrated from the circuit for a synchronous counter given in Figure 27, where an asynchronous reset has been applied. Let us assume that this circuit has been used to provide signals for other parts of a logic network and, due to the glitch inherent in the design, is malfunctioning. We can connect a logic timing analyser to the clock line and to the three bistable outputs Q_A, Q_B, and Q_C. The clock used to update the counter runs, say, at 2 Hz, giving a clock period of 500 millseconds. The complete cycle of the counter will have to be viewed for 2.5 seconds to capture and display all the counter states. Let us assume that our analyser has a RAM which is 256 bits deep, so that by setting its sampling period at 100 milliseconds one complete counter cycle will just underfill the storage RAM. As with logic state analysers, logic timing analysers can be qualified to begin capturing information only when the preset triggering criteria have been met. Let us set the analyser to begin capturing data when the three counter outputs are all at logic "0" level. On a normal oscilloscope, the reset glitch on the Q_A output will be virtually undetectable because of the slow timebase speed set to ensure that all the counter states can be viewed. The glitch itself is only approximately 68 nanoseconds wide and will pass unnoticed on a trace that takes 2.5 seconds to cross the oscilloscope faceplate. A logic timing analyser, however, will detect the reset pulse on the Q_A output as being greater than 5 nanoseconds in width but less than 100 milliseconds and display it as a brightened up glitch signal.

On the majority of timing analysers, the minimum glitch width can be preset so that signals greater than 50 nanoseconds but less than the sampling period can be detected. Data capture can also be qualified as a combination of events, including glitch occurrence, on one or more of the input lines. This particular feature is very useful where an event only happens intermittently and frees the user to carry out other tasks, while the analyser is left to wait for the event to occur. Timing analysers, in common with state analysers, may be set up to capture pretrigger as well as delayed events, which in combination with a glitch triggering capability provides for a very powerful analysis tool. Thus the events that lead up to the timing error can be captured, stored and displayed for analysis.

A useful feature of timing analysers is a moving cursor which, as it is moved along the timing displays, provides an on-screen display of delay time relative to the trigger pattern at the beginning. Using this facility, the operator can measure the time between events within the limits of the internal clock. Thus if the clock runs at 20 Mhz, relative time measurements to a resolution of ± 50 nanoseconds are possible. The actual accuracy to which this relative time measurement can be made, however, depends upon the number of digits used to represent it on the screen. If only 3 digits are used, and the time measurement is 15.648 milliseconds, then only 15.6 milliseconds will be displayed. For most practical purposes, with the exception of parametric analysis where a precise value may be needed, this accuracy is sufficient for most applications.

6.3 DISPLAY MODES

Information may be displayed on a logic analyser in a number of different ways, each of which has advantages and disadvantages. These display modes have evolved to meet a demand as the instruments themselves have developed. The earlier, and supposedly simpler, display forms may reveal a fault which may be clouded when the information is presented in a more comprehensive form. Take as an example the following sequence:

1101 0111	D7
1011 0000	B0
0100 0100	44

The information on the left-hand side is presented in binary and that on the right-hand side in hexadecimal. This can represent the 8 lines of a data bus in which, through a fault, the line D_3 is being held permanently low. This is apparent from the simpler binary display, but is not clear from the hexadecimal codes.

Currently these are 7 different display modes that are used in logic analysers, although no one instrument is capable of displaying them all.

6.3.1 The MAP mode

Consider that we have just designed and constructed a microcomputer system and have written a short test routine to exercise the

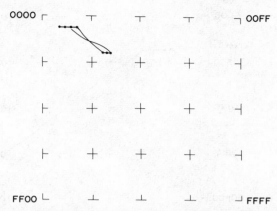

FIGURE 54   MAP DISPLAY QUADRANTS IN HEXADECIMAL NOTATION

machine. When we first switch on we are primarily interested in whether the machine accesses the addresses of our test routine or, due to an error, attempts to execute a non-existent program elsewhere in the memory. To determine the activity of the system entails only viewing the states of the address bus lines, because at this point we are only concerned with their activity, regardless of any associated data bus transfers. Most early logic analysers, and some more recent models, allow the user to view address bus activity using the MAP mode of operation. The MAP mode displays logic words up to 16 bits long as a single dot on the screen with each representing a unique word. The whole screen may be used to represent the 65,536 possible address values of a 16-bit bus with 0000 as a dot in the upper left-hand corner of the screen, down to FFFF in the bottom right-hand corner. When the SUT uses any address, it appears as a bright dot at the appropriate position. By incorporating a trace between the dots, address bus activity can be followed from the screen map. On the Hewlett Packard model 1607S analyser, the trace between dots is wider at the end nearest the first address in a sequence and narrows to the next. Figure 54 illustrates the manner in which the screen is divided to represent the entire directly addressable space of an 8-bit microcomputer using a 16-bit address bus.

The activity is meant to represent the type of picture that will result from a routine running in low order memory which calls a subroutine in higher order address locations. This can, for example, be the type of activity that will occur in a microcomputer which is repeatedly reading a keyboard for an entry and which remains in a loop until a key is pressed. To determine an address value precisely, the switches that will normally be used to set up a qualifying trigger address are used to position a cursor on the MAP display and the value read from

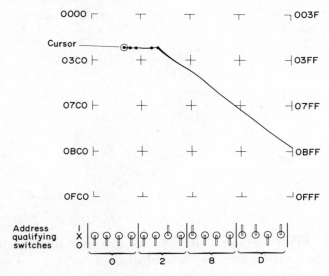

FIGURE 55   EXPANDED MAP MODE COVERING ONE OF THE 16 QUADRANTS

the switch settings is then the displayed address ringed by the cursor. The display as illustrated by Figure 54 is too coarse for exact address location; each of the 16 display quadrants may be expanded to fill the screen, so that any address may be precisely located with the cursor. Each quadrant only represents 4096 address locations, each of which is separable on the whole screen using the cursor. Typically, using a magnification control, the quadrant where the cursor is located may be expanded to fill the whole screen; the cursor is then moved about until it covers the required dot and its address is read from the qualifying switch pattern. Figure 55 illustrates an example where low order address bus activity is taking place and the cursor has been positioned over one dot to discover its precise value.

From Figure 55 the controls give a value of 028D for the address of the displayed dot covered by the cursor. The expanded quadrant illustrated is the first in the total address space, which covers the first 4096 address locations, i.e. 0000 to 0FFF in hexadecimal.

The MAP mode of display is particularly useful when developing a system to give a qualitative view of the complete system behaviour and will show up any anomalies. Many manufacturers of logic analysers have, however, dispensed with it on their later and supposedly more sophisticated machines.

On a MAP, the position of a dot indicates its binary magnitude, and its brightness defines its relative frequency of occurrence. The trace between two dots signifies the binary pattern being changed from one

FIGURE 56   6-STATE COUNTER MAP DISPLAY (GREATLY EXAGGERATED)

value to another, usually with a thicker trace at the emanating dot of the vector and a thinner trace at the destination dot. The MAP is essentially a pictorial method of presenting binary activity in a system and does not have to be restricted to viewing address activity alone. One can view the activity of, say, a binary counter using the MAP mode and the resulting display will be a unique pattern for the counter.

Figure 56 depicts the display that will result from viewing the outputs from a 6-state binary counter which cycles through its 6 possible states and then resets back to its zero state. The display is unique to this counter and can be considered as a "fingerprint" for any such binary counter. One can with practice learn to recognise characteristic patterns from MAP displays and identify them with particular types of circuit or system behaviour. MAP displays are not often used because they present address bus behaviour in an unfamiliar way, which has to be learnt if any meaning is to be attached to it. For rapid identification of address bus activity, as a first glimpse into what a system is doing, the MAP mode is hard to better. Logic analysers such as the Hewlett Packard 1607S, Tektronix 7D01, and Gould Biomation model 1650-D, all include the MAP mode within their repertoires.

6.3.2 State flow-binary and grouped binary

As soon as logic analysers began to use internal RAM to store the data captured from a SUT, then it became natural to present that data in a binary format on the display device. Each address and corresponding data value was displayed as a series of logic "0"s and "1"s across the screen, with the trigger word and any pretrigger values towards the top of the screen and delayed values below them. To ease the readability of the displayed information, the binary digits were grouped into threes or fours and read as octal or hexadecimal values. These types of display relate to logic state analysers and indicated the state flow within a system; a typical display was as follows (Figure 57).

The left-hand bank of binary digits represented the state flow captured from some system and was grouped into 4s to enable the operator to translate the display into hexadecimal values. For readability, the display was also split vertically into groups of 4 lines of

```
0101 0111 1101 0100          0101 0111 1101 0100
0010 1011 1010 1000          0110 0011 1000 1100
0001 0100 0101 0000          0011 0000 0001 1000
1100 0100 0100 0110          0100 1100 0110 0100

0110 0011 1000 1100          1010 0111 1100 1010
0011 0000 0001 1000          1101 0011 1001 0110
1001 1000 0011 0010          1110 1000 0010 1110
0100 1100 0110 0100          0111 0100 0101 1100

1010 0111 1100 1010          1011 1011 1011 1010
1101 0011 1001 0110          1101 1100 0111 0110
1110 1000 0010 1110          0110 1111 1110 1100
0111 0100 0101 1100          0011 0111 1101 1000

1011 1011 1011 1010          0001 1011 1011 0000
1101 1100 0111 0110          0000 1100 0110 0000
0110 1111 1110 1100          1000 0111 1100 0010
0011 0111 1101 1000          1100 0011 1000 0110
```

FIGURE 57   GROUPED BINARY STATE FLOW DISPLAY

information. The top line of the display will translate to 57D4 in hexadecimal and can represent an address captured from a microprocessor system.

Taken as a whole, each line in the display occupies both the left-hand and right-hand banks and will represent 32-bit words which can be composed of a 16-bit address, an 8-bit data bus value and a further 8 external lines. In this instance the left-hand side will represent address bus information while the right-hand side will be data bus values and external line states.

Most analysers that use this mode of display can, however, have the screen split into two separate portions where the right-hand side will be data collected from a SUT at some time while the left-hand bank will represent data captured from the same lines at a later time, using the same trigger qualifiers. Where any differences occur, the differing digit is brightened up to indicate a change of state from the stored bank of data on the right-hand side. Figure 57 indicates just such a set of events, where the right-hand bank is data captured from a system and the left-hand bank represents data from the same source some time later. The overprinted digits are meant to indicate those bits that differ from one bank to another. This form of display is most useful where inputs are being tested from a set of digital transducers such as limit switches or proximity detectors. Here we are primarily interested in the binary states of the signals, and the binary display, whether grouped or not, is ideally suited to our purpose. Input state changes can be readily seen from the brightened up display of Fig. 57 to indicate that limit switches or similar devices have been changed between sampling and storing the first set of readings and taking the second set.

As previously stated, binary type displays are useful for checking for lines that are shorted to ground (permanently at "0"), or tied to the positive supply rail (permanently at "1"). In this sense they act as a logic probe, capable of storing many past events, rather than just displaying the current event being viewed.

Logic analysers that permit this comparison of two sets of data and, by Exclusive ORing them, produce bright-up on bits that differ, include Gould Biomation's model LA5000, Hewlett Packard's 1607S, and Tektronix 7D01 with a DF1 display formatter.

6.3.3 State flow-hexadecimal format

The next evolutionary step in displaying state flow in a logic state analyser was to convert the grouped binary digits into octal or hexadecimal characters. While octal was primarily used in the mini-computer era, the majority of microprocessor systems use hexadecimal coding. A hexadecimal formatted display is of use where there is a machine code listing of the program which is to be checked and it is possible to obtain a one to one correspondence between the screen display and the listing.

As an example, assuming that a delay subroutine had been written, beginning at address 0F00 and that the assembly listing was as follows:

LABEL	OPCODE/OPERAND	COMMENT
DELAY:	LD B,FF	;load register B with 255_{DEC}
DEL:	LD C,FF	;load register C with 255_{DEC}
LOOP:	DEC C	;reduce register C by one
	JP NZ,LOOP	;if register C$\neq$0, jump to LOOP
	DEC B	;reduce register B by one
	JP NZ,DEL	;if register B$\neq$0, jump to DEL
	RET	;if register B=0, return.

This delay subroutine, written in Z80 mnemonics, is typical of the type of program used to provide a short delay. When assembled into machine code, starting at address 0F00, the program will look like:

ADDRESS	OPCODE/OPERAND
0F00	06 FF
0F02	OE FF
0F04	OD
0F05	C2 04 0F
0F08	05
0F09	C2 02 0F
0F0C	C9

When this program is called from a main program, the inner loop labelled LOOP will be executed 255 times until register C becomes zero. The outer loop labelled DEL is then executed, which decrements register B and calls the inner loop for another 255 cycles. The effect is to call the inner loop 255, which itself executes 255 times. The overall effect is to call the inner loop something over 65,000 times and for a Z80 using a 4-MHz system clock will provide a delay of approximately 230 milliseconds.

If we trigger the logic analyser at address 0F00, then when the subroutine is called the following display will result:

ADDRESS	OPCODE/OPERAND	
0F00	06	
0F01	FF	
0F02	0E	
0F03	FF	
0F04	0D	
0F05	C2	
0F06	04	
0F07	0F	
0F04	0D	
0F05	C2	
0F06	04	
0F07	0F	
0F04	0D	ETC

What we are actually seeing is each separate opcode or operand of every instruction transferring over the system busses. When the CPU receives the complete conditional jump instruction at location 0F07, the decrement on the C register has not yet made it zero, and so the program loops back to location 0F04 to execute the decrement C instruction again. The limited capacity of the analyser storage RAM means that we cannot possibly store and view the complete subroutine unless the initial numbers stored in the B and C registers are relatively small. We can retrigger the analyser on address 0F08 to ensure that it completes the inner loop when register C is decremented to zero, and we can further trigger the analyser on the return instruction at location 0F0C and dial in some pretrigger value to view the completion of the subroutine.

The displayed information is in agreement with our machine code listing and can clearly be used to verify that the program has been entered correctly and excutes properly. Without further qualifying features, however, this type of program checking can be extremely tedious and time-consuming. Other features generally associated with qualifying trigger words which simplify this problem of testing a loop program, that may take many cycles to complete, are discussed in Section 6.4 under Typical Analyser Features.

6.3.4 State flow-disassembled format

Currently the most sophisticated display format for state flow within a sequential machine is to present the information in disassembled form. Clearly, because each microprocessor uses different mnemonics for its instructions, a machine has to be fitted with a personality card to suit the processor system under test. For the example given in Section 6.3.3, the equivalent disassembled display will be:

ADDRESS	OPCODE/OPERAND	EXTERNAL
0F00	LD B,FF	0000 0000
0F02	LD C,FF	0000 0000
0F04	DEC C	0000 0000
0F05	JP NZ, 0F04	0000 0000
0F04	DEC C	0000 0000
0F05	JP NZ, 0F04	0000 0000
0F04	DEC C	0000 0000
ETC		

The display in terms of program flow is now in a more readable form, with the mnemonic instructions being displayed in place of their machine code equivalents. Obviously the analyser cannot interpret labels which were used in the preassembled form of the source program, and can only substitute the actual address values when conditional jumps and calls are encountered.

Although less prone to error in interpreting the display, the same problems of needing to further qualify the display to check the overall timing and number of times a loop is executed still apply.

An analyser capable of being personalised to suit a variety of microprocessors and which can display disassembled code is the Hewlett Packard model 1611A.

6.3.5 Timing displays

While some analysers such as the Gould Biomation models 9100-D and 920-D are only meant to display timing waveforms for parametric and functional timing analysis, the majority of analysers that have timing display features will also present grouped binary, hexadecimal and map mode displays. The Gould Biomation model 1650-D, for example, will display data as up-to 16 channels of timing, in grouped binary with associated hexadecimal translation, or in a MAP mode.

The Hewlett Packard model 1615A analyser will provide state flow displays in either binary, octal, decimal or hexadecimal characters and can also be used as an 8-channel functional timing analyser.

Timing displays play an important role when a timing problem exists, and are usually associated with the use of control bus lines and/or external inputs. Glitches that cause phantom interrupts can, for example, be captured and subsequently displayed on a timing analyser, even when they only occur intermittently.

6.3.6 Special coding formats

If one has to analyse an asynchronous serial data transmission line connecting a computer to a peripheral such as a printer or VDU, then data will be transferred one character at a time in a bit serial manner. So that more than just numeric data can be transferred over the line, a coding technique is used where the character bit pattern defines it as an alphanumeric character or a control word. The commonest coding method in use is the *American Standard Code for Information Interchange* or ASCII for short. The majority of analysers we have considered so far have been designed to handle data in a parallel format, while here we are faced with bit serial information. Special analysers have been developed which will decode the serial bit patterns and translate them into their equivalent ASCII words for subsequent display. The commonest electrical and mechanical standard for a serial interface is the EIA RS232C standard, and analysers such as Hewlett Packards model 1640B are capable of measuring and interacting with such serial links so that they may be fully exercised and tested.

The same type of problem exists when testing an IEEE-488 parallel interface in which 16 lines are used to communicate between systems. All address and data information is trafficked over 8 lines with the remaining 8 acting as control lines. All characters are normally sent over the 8-line bus in ASCII code, while each transfer is controlled by 3 handshake lines. The remaining 5 lines are used to identify the data on the bus as either address or data information and to provide general managerial supervision to the system. Special mnemonic codes relate to the bus transactions, which can be analysed and displayed in their mnemonic formats using analysers such as Hewlett Packard model 1602A, or Tektronix 7D01 analyser with special DF2 display formatter.

6.4 LOGIC ANALYSER FEATURES

The early logic analysers were hardwired systems that had to be programmed manually from front panel controls to set up qualifying parameters such as trigger address and data bus values. The displays were limited to grouped binary or timing waveforms and could not be said to be well designed ergonomically. With the introduction of a microprocessor as the control element in an analyser, there came the possibility of prompting the user for input set-up data through the use of screen layouts which requested trace specifications such as address bus qualifying states, etc. Many logic analysers now tend to make use of the "menu" concept, where several screen formats are used that require the user to enter the information needed by the analyser before it can begin data capture.

TRIGGER	ADDRESS 0F00	DATA XX	EXTERNAL XXXX XXXX

— — — — — — — — — — — — — — — LINE 0 — — — —

ADRS	OPCODE/DATA	EXTERNAL
0F00	LD B,FF	0000 0000
0F02	LD C,FF	0000 0000
0F04	DEC C	0000 0000
0F05	JP NZ,0F04	0000 0000
0F04	DEC C	0000 0000
0F05	JP NZ,0F04	0000 0000
0F04	DEC C	0000 0000

This display is the type that will result on a Hewlett Packard 1611A logic state analyser with a personality module fitted to disassemble Zilog Z80 mnemonics which have been triggered on address 0F00 and where "don't care" states have been entered for the data bus value at address 0F00 and the 8 external lines are not used as trigger qualifiers. The 1611A analyser has a set of keypads through which set-up parameters are entered, and which also allow the user to begin tracing state flow in a system. The line drawn around the first part of the captured data is meant to represent the fact that it will normally be displayed in reverse video to indicate that it is the trigger word.

If any pretrigger requirements have been keyed into the system, then above the line number (LINE 0) there will appear the number of events requested. The display will show these events while still maintaining the trigger word in reverse video. The trigger word will thus appear lower down the screen, following the number of pre-trigger events. The line number shown indicates that the top line of the display is the trigger word, and because the 1611A is limited to displaying only 16 lines at a time, scroll keys permits the user to roll up the display to view all 64 captured events. As the lines are rolled up, so the line number increases to indicate that the top line of the display occurs later in the data store after the trigger word.

6.4.1 Counting events

When debugging software, we are sometimes concerned to find out how many times a particular event has been executed. An example is to determine how many times the decrement C register (DEC C) instruction has actually been executed when the complete delay subroutine has been called. Many analysers allow the number of trigger events that occur between Enable and Disable qualifying words to be counted. Thus, for example, if an analyser is enabled on address 0F00 and disabled on address 0F0C and the trigger event has been set to address 0F04, then between the first and last addresses of the delay subroutine the number of times that address 0F04 has been used will be counted by the analyser and displayed. To accomplish this type of measurement, the analyser's keypad has several sets of trigger keys labelled TRIGGER ENABLE, TRIGGER and TRIGGER DISABLE. Each set of keys can be used to specify address bus, data bus and external line states as qualifiers or have "don't care" states entered in those parameters that will not take part in trigger qualification. Usually, only address bus information is used.

	ADDRESS	DATA	EXTERNAL
ENABLE	0F00		
TRIGGER	0F04		
DISABLE	0F0C		

— — — — — — — — — — — — — — — STOPPED — — —

ADRS	OPCODE/DATA	EXTERNAL

COUNT	65025 TRIGGERS
MAX	65025 TRIGGERS
MIN	65025 TRIGGERS

The qualifying set-up parameters are indicated above the dotted line which also shows that the analyser has captured the required information and stopped. The analyser was enabled at address 0F00 and "don't care" states were entered for data bus and external line values. Typically, these are then not displayed, and if no information is entered for the data bus and external lines then the analyser defaults to "don't care" states. This enable address corresponds to the first address of the subroutine and in a similar manner, using the disable keys, a disable address of 0F0C was entered. This corresponds to the final RETURN instruction in the delay subroutine. The analyser can now be used to count the number of times that a qualifying word occurs between the address limits of 0F00 and 0F0C. The DEC C instruction is located at address 0F04, which is used to set up the trigger word for the analyser using the trigger word keys on its keypad. On the 1611A, a separate COUNT TRIGS key is pressed to initiate event counting, which will begin when the enable address is first encountered and terminates when the disable address is captured.

The final display is as shown above, which indicates that the DEC C instruction is executed 65,025 time during the delay subroutine execution time. Beneath the count, there are two further figures displayed which indicate the maximum and minimum number of trigger events captured. Using the COUNT TRIGS key, the number of trigger events can be counted for several calls of the subroutine and any differences displayed on the MAX and MIN lines. The MAX and MIN values are of interest where a branch may occur between the enable and disable trigger words which may bypass the actual trigger event; this will result in different counts, depending on the action of the branching instruction.

6.4.2 Time interval measurements on a Logic State Analyser

The execution times of certain parts of a program may need to be known to a reasonable degree of accuracy. An example is the time taken to execute a delay subroutine which, due to a slight difference between the actual system clock frequency and its nominal value, can cause a difference between the actual and expected delay times. The delay subroutine could be used to allow time for an analogue to digital converter to carry out a conversion on an analogue input signal and which for a device such as the AD574 converter would require a delay of 25 microseconds. The converter would be sent a convert command and then a 25-microsecond delay subroutine called to allow for the convert time before reading the digital value from the converter. If, due to a difference between the nominal value and actual value of the system clock frequency, the delay happened to be only 20 microseconds, then when read the converter would provide a nonsensical result. Another example of time interval measurement is the estimation of the time taken between two external events which, due to the critical nature of a process, have to occur within some predefined time limit. The events may correspond to operating an actuator and sensing its response from a proximity device. The processor may be put into a wait loop, awaiting the setting of the proximity detector, before the system is allowed to proceed. A watchdog timer may be operated concurrently with the wait loop to ensure that a response happens within some predefined time limit; the watchdog timer in microprocessor systems is usually a *Counter-Timer Circuit* (CTC) device which is programmed to provide a fixed time delay and interrupt the CPU when that time elapses. If the proximity detector responds before the CTC time elapses, the CTC is disabled and the program continues. In many systems where mechanical actuators are used, the maximum time that has to be allowed for the operation is unknown, which prevents the CTC from having an accurate time value loaded into it. By making time interval measurements on a system, the maximum time that has to be allowed for the operation can be measured and used as a time constant for the CTC. If then the operation takes longer than the maximum time, the CTC interrupts the CPU to indicate a mechanical fault in the system.

Time interval measurements are set up on a logic state analyser using the enable and disable trigger words with the actual trigger event left as "don't care" states. A TIME INTVL key is used to initiate the timing measurement which commences when the system inputs agree with the parameters set up on the enable keys and stops when the disable trigger conditions are satisfied. The accuracy of the measurement is usually defined by a clock running inside the analyser and will typically be to within 1 microsecond.

	ADDRESS	DATA	EXTERNAL
ENABLE	0F00		
TRIGGER	XXXX		
DISABLE	0F0C		

— — — — — — — — — — — — — — —   STOPPED — — — —

ADRS	OPCODE/DATA		EXTERNAL

TIME	228928 MICROSECONDS
MAX	228928 MICROSECONDS
MIN	228928 MICROSECONDS

The above example of a time interval measurement on the delay subroutine has assumed a clock period of 250 nanoseconds, which is typical for a Z80A microprocessor running on a 4-MHz clock. The time taken to execute the subroutine has been measured at 0.229 second and in common with the count triggers mode of operation, maximum and minimum times are also displayed in the event of there being a difference.

In the delay subroutine example there is unlikely to be any significant difference in its execution time, but for the example of a system waiting for a response from an external input the time difference could be substantial.

```
ACTUATE:     LD  A,08          ; set actuator bit in output port
             OUT(04),A         ; whose I/O address is port 4.
INPUT:       IN  A,(02)        ; read input port to test proximity
             AND 01            ; device tied to bit 0 of port 2.
             JP NZ,INPUT       ; jump around INPUT loop until bit
                               ; 0 sets to "1".
```

This Z80 assembler routine sets an actuator connected to output port 4 and then waits in an input read loop until a proximity detector or limit switch attached to the least significant bit of input port 2 sets. In practice, prior to this routine, a CTC will be programmed to time the operation and interrupt the CPU if the set time elapses. If an input occurs before the set time, the subsequent instructions in the program will disable the CTC to prevent it from interrupting the processor.

The analyser can be enabled on the address value associated with the label ACTUATE and disabled on the address of the instruction following the *Jump if Not Zero* (JP NZ,INPUT) conditional test. Several time measurements will be taken to estimate the time needed to be set into the CTC for optimal detection of a failure in the system.

6.4.3 Trigger ranges and trigger occurrence

On many analysers, rather than specify one single trigger pattern, the triggering event may be qualified as an address value lying within a range of addresses. We may desire to capture any events that occur between address 0100 and address 0200, and these can be keyed into the analyser to provide a trigger window. The first address that is captured by the analyser and lies between these values acts as the trigger word for the instrument.

Facilities are often included in an analyser which allow the preset trigger word to occur a number of times before any information from the SUT is captured and stored. This feature is used when the number of passes through a loop are known, but we are only interested in displaying the last pass and subsequent program instructions. Say, for example, than a loop is executed 20 times before proceeding onto the next portion of a program. We can key in a trigger occurrence value of 19 and then start the analyser. Information will only be captured from the SUT after 19 occurrences of the trigger word, which will cause only the final pass through the loop to be displayed along with subsequent instructions in the program.

6.4.4 Pre- and post-triggering

Under normal circumstances, a logic state analyser will capture and display an input pattern that corresponds to its present trigger word and subsequent events up to the limit of its internal storage. Many analysers are limited to storing only 64 events and can thus only present to the user a small amount of information about system activity at any time. To extend the capabilities of an analyser, a number of pretrigger events can be specified, up to the limit defined by the size of the internal storage. Thus for an analyser that can store 64 events, up to 63 pretrigger events which occurred prior to the trigger word can be stored and displayed. This feature is clearly useful in fault-finding, where an occasional erroneous event take place and we wish to trace the events that led up to it.

Post-triggering introduces delay into the measurements by allowing the user to specify the number of system transactions that can occur following the reception of the trigger pattern by the analyser, but before any information is captured in the instrument. In the delay subroutine example we can specify a delay value of 65024 and set the trigger word address at 0F04, which will allow the delay subroutine to execute all but its last cycle before being stored and displayed on the screen. This appears at first sight to be similar to specifying the number of trigger occurrences that can occur before data capture is initiated; the delay value is usually, however, far greater than the number that can be specified, and thus provides a greater scope for effective digital delay.

6.4.5 Personality modules

For specific microprocessors, some logic state analysers can be fitted with personality modules which allow disassembled code to be displayed using the particular microprocessor mnemonic codes, and also allow the user to view the more important control bus signals. Extra trigger qualifiers fitted onto the personality module can also be used to specify the types of bus transfers that are stored and displayed. One may, for example, only by concerned with I/O transfers and use a facility on the personality module so that only I/O transactions are stored and displayed on the analyser screen. The bus transactions may be further qualified to be only write operations which, in conjunction with an I/O only qualifier, would cause only writes to output ports to be viewed on the analyser.

The major control bus signals are usually displayed as LED indications on the module, so that the user can see at glance if any fault condition on a control line is preventing the system from functioning correctly. Typically, the absence of a clock will be indicated along with such lines as RESET and WAIT which, if active, will prevent the processor from executing any programs.

A logic state analyser fitted with a personality module for the microprocessor system under test is currently the quickest and easiest way to debug the software of a system. Provided one is familiar with the instrument and is conversant with its display format, many I/O problems can also be resolved using a personalised analyser.

6.4.6 Qualifiers on Logic Timing Analysers

Logic timing analysers will usually display 4, 8 or 16 timing traces concurrently. In common with logic state analysers, timing analysers allow for a flexible range of trigger conditions before beginning to capture data from a system. Typically, the state of every input line can act as a qualifier and/or a glitch on any number of input lines can be used to extend the trigger specification.

The trade-off between depth of storage and clock rate means that a timing analyser must be capable of being set up over a wide clock range from 20 nanoseconds up to 0.5 second.

A timing analyser should be able to provide on-screen timing measurement using a movable cursor so that accurate time differences can be measured from the display. Many analysers also allow part of the display to be expanded for greater timing accuracy.

6.4.7 Accessories

Leads have to be connected to the SUT to monitor the bus transactions that are taking place. A general purpose analyser will be provided with a number of pods to collect data from the SUT, each of which will typically have up to 9 leads connecting into it. There may be up to 4 pods in use to collect information from the system busses and external lines. Each line needs to be wired into the SUT, and the normal method is to use "grabbers" which are small clips which can be connected onto one pin of an IC package without shorting out to neighbouring pins. Where one is monitoring activity on 32 lines, it is a tedious and difficult task to connect up all the "grabbers".

For a general purpose analyser, one effective solution is to use an

IC clip onto the microprocessor and bring all its pins out to accessible points on top. For a specific system, a clip may be itself hard wired by dispensing with the "grabbers" and soldering the analyser input leads directly onto the clip to provide a test harness.

For personalised analysers, the microprocessor is usually removed from the SUT and replaced with a DIL header which connects into a pod into which the microprocessor is then plugged using a *Zero Insertion Force* (ZIF) socket. With the exception of external lines, all the necessary analyser inputs can be taken from the microprocessor which eliminates the need for many flying leads connecting the analyser to the SUT. One major advantage of placing the SUT microprocessor in a buffered pod is that the analyser can be used to control as well as monitor the SUT. By forcing a HOLD state onto the processor from the analyser, a program can be single stepped through from the analyser which may not be feasible using the SUT alone.

Chapter 7

Signature Analysis

Analogue circuits are tested by stimulating them with known test patterns and checking each node in the signal propagation path with an oscilloscope. If and when an unexpected waveform is detected, then the part of the circuit driving the node is suspect and investigated. In an analogous manner, digital circuits may also be stimulated with known test patterns and each node in the signal propagation path tested. Digital systems, however, differ radically from analogue systems, not only due to the basic nature of the signals involved, but usually a digital system will have many more signal inputs. If a digital system can be controlled so that all its inputs form a well-defined test pattern, then each and every node will be exercised through a fixed pattern of events. For any node, the same pattern will occur every time the stimulation program is run, and this can be used to verify the correctness of the node. If the measured pattern differs from that expected, then a fault can be inferred between it and the stimulation points of the system. Given a fixed test pattern, then for any node the pattern measured will be unique and can be used to check it on a go/no go basis. The term "signature" has been applied to the measured response of a node to a known test pattern by analogy with the uniqueness of a person's handwritten signature. This concept has led to an area of digital testing called *Signature Analysis*.

Signature analysis developed from two previously used error checking techniques; the first was an already established method of testing logic nodes in a digital network called transition counting, and the second was an error checking technique called *Cyclic Redundancy Checking* (CRC). Signature analysis emulates the former, while using a method similar to the latter.

Both transition counting and signature analysis are tests on the hardware of a system, and to understand why and how they came into being we need to review the types of waveform that can be found in a digital system, including computer based schemes.

131

7.1 THE NATURE OF DIGITAL WAVEFORMS

The type of signal used to clock both synchronous and asynchronous digital networks is repetitive and can be characterised by its pulse repetition frequency and its duty cycle or mark/space ratio. Duty cycle or mark/space ratio is a measure of the time a digital waveform spends in the logic "1" state compared to the time it spends in the logic "0" state. If a signal spends as much time in the logic "1" state as it does in the logic "0" state, then its duty cycle is said to be 50 per cent or it is said to have a 1:1 mark/space ratio. The pulse repetition frequency of a digital signal is the inverse of the time between successive positive going transitions (or negative going transitions) and only applies to repetitive waveforms.

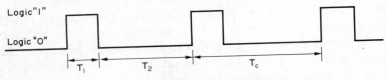

FIGURE 58   A PERIODIC DIGITAL WAVEFORM

The periodic digital waveform shown in Figure 58 has a pulse repetition frequency (prf) given by $1/T_c$. The mark or on time (T_1) plus the space time or off time (T_2) are equal to the period T_c. The waveform is characterised by the two parameters of prf and mark/space ratio, both of which can be measured using a conventional oscilloscope. About the only signal that corresponds to the regular type of waveform shown in Figure 58 in a digital computer system is the clock.

FIGURE 59   A NON-PERIODIC DIGITAL WAVEFORM

Within a digital computer or any complex logic network, the majority of waveforms appear to be almost random when viewed on an oscilloscope. A signal may appear as depicted in Figure 59 where the mark times and the space times vary considerably over the trace length. No duty cycle can be defined for the signal, nor can any pulse repetition frequency be established to characterise it. The non-stationary nature of these signals means that no statistical method can be applied to identify them. The reasons for the irregular nature of digital signals in complex logic networks are as follows.

Any single node in a logic network that operates on data in a parallel format provides only part of the information present within the system at any instant. The piece of information will change apparently randomly with time as the characters being transferred within the system change. A simple analogy to the problem would be to take one single alphabetic character from each word on this page at random, and then try to make sense of the information, using only the very limited subset of information collected.

Programs running in a computer will change due to conditional branch instructions, calls to subroutines, interrupt requests, and *Direct Memory Access* (DMA) operations. These may be viewed as either software initiated or hardware initiated asynchronous events which have a randomising influence on the logic waveforms. Under normal operating conditions, characterisation of the waveform present on a single node is impossible due to effects of unexpected events.

7.2 TRANSITION COUNTING

If a node can be stimulated with a known and repeatable test pattern, then in effect we can treat the nodal signal as a pseudo-random binary sequence which in the short term may appear random, yet repeats over a longer time scale.

One established technique is simply to count the transitions of the signal from one logic state to the other and use the final count value as an identifier for the node. The total number of state transitions for a stimulated node may be many thousands, and so some form of compression technique has to be adopted. A necessary condition for transition counting is a defined "time window" only during which transitions on the node are counted. Within this "time window" a test program is run which exercises the node under test and which is preferably repeatable so that several measurements can be made and correlated. For a given system and a defined "time window" and test sequence, the results of several transition counts should be identical. To use transition counting as a fault-finding technique, a working system is taken and the *Transition Count* for every node is measured and documented. When a fault develops, the tester runs the test programs, measures the count from suspect nodes and compares them with their documented values. Any discrepancies indicate a fault and by working systematically through the system it can be isolated.

7.3 THE PROBABILITY OF SUCCESS USING THE TRANSITION COUNTING TECHNIQUE

The implementation of transition counting as a testing method must employ some form of instrument which between defined start and stop signals (the "time window") will count every state transition on a network node. In common with any other electronic system, the transition counter is itself prone to error by miscounting. It is being used as a measure of goodness of the *System Under Test* (SUT) and so we are interested in how well it can perform its task. The length of any test sequence is usually unknown, as are the number of state transitions within it; for this reason we can only predict a probability of success for transition counting as a general technique.

For the simple, illustrative example given in Figure 60, within the time window there are 9 transitions, excluding the initial state change on the data line that is coincident with the leading, active edge of the start signal.

In order to calculate the probability of failure of a transition counting instrument, we will first assume that we have an m-bit binary sequence, in which there are r state transitions. In general, $r \neq 2m$, because a bit period will be defined by some system clock signal and the data being analysed will be in a *Non-Return to Zero* (NRZ) format, as illustrated in Figure 60.

For binary signals, there are 2^m possible m-bit sequences that can exist, and we will define the total number of m-bit binary sequences as:

$$N_s = 2^m.$$

The transition counter has no memory of the correct sequence that could occur at any node. So, if an error does occur, it cannot relate its position to the proper sequence in time. The order in which transi-

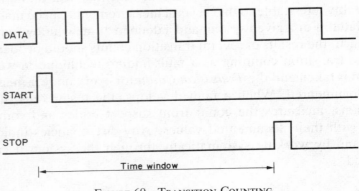

FIGURE 60   TRANSITION COUNTING

tions are counted is therefore irrelevant, and the possible number of ways of counting r transitions in an m-bit sequence is given by the Combinations rule:

$$N_r = \binom{m}{r} = \frac{m!}{r!(m-r)!}.$$

We will further assume that in a system N_t errors occur, and that N_u of those errors go undetected. The probability of a transition counter failing to detect an error, assuming that some errors do occur, is:

$$\text{Prob(TC,fail)} = \frac{N_u}{N_t}.$$

But

$$N_u = \sum_{r=0}^{m} P_{ur},$$

where P_{ur} is the probability of undetected errors given r transitions.

$$P_{ur} = N_{ur}P_r$$

where N_{ur} is the number of undetected errors given r transitions, and P_r is the probability of counting r transitions. Now

$$N_{ur} = N_r - N_c$$

where N_c is the number of correct ways of counting and can only be equal to one ($N_c = 1$). Also

$$P_r = \frac{N_r}{N_s}.$$

We can now rewrite

$$N_u = \sum_{r=0}^{m} (N_r - 1)\frac{N_r}{N_s}$$

and

$$N_t = N_s - N_c = N_s - 1.$$

Hence,

$$\text{Prob(TC,fail)} = \sum_{r=0}^{m} \frac{(N_r - 1)(N_r/N_s)}{2^m - 1}.$$

This is the probability of a transition counter failing to detect an error, and so the probability of the counter detecting an error is

$$P(TC) = 1 - \text{Prob(TC,fail)}.$$

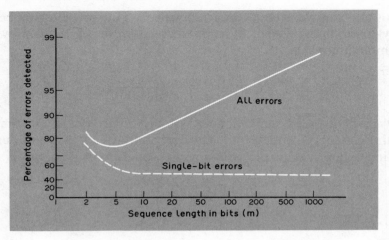

FIGURE 61   THE PERCENTAGE OF ERRORS DETECTED

Expressed as a percentage and expanding for N_r, the probability that a transition counter will detect an error in a sequence of length m is given by:

$$P(TC\%) = 100 - \left(\frac{100}{2^m(2^m-1)}\right) \sum_{r=0}^{m} \left[\frac{m!}{r!(m-r)!}\right]\left[\frac{m!}{r!(m-r)!} - 1\right]$$

$$\approx 100\left(1 - \frac{1}{\sqrt{m\pi}}\right).$$

We can plot this equation for increasing m as a graphical method of assessing the technique.

This equation is really only valid where errors affect the entire m-bit sequence, which is a situation that rarely arises in practice. usually errors are more subtle and affect a few bits, or in the worst case only a single bit.

There are 2^m possible m-bit sequences and any bit can be changed to produce a single bit error, so there are $m \times 2^m$ possible single-bit errors.

We will assume that the initial state of the transition counter is zero. If the second bit in the m-bit sequence is a "1", then the state of the first may be changed and will go undetected. Under these conditions, there are 2^{m-1} ways of completing the sequence.

From Figure 62 it can be seen that if the first bit of a sequence changes when the second bit is a logic "1", then the transition counter still only counts one state change even when an error is present.

The second possible way that a single-bit error can go undetected

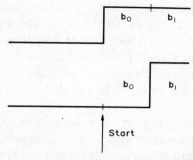

FIGURE 62   UNDETECTABLE FIRST BIT STATE CHANGE

in a transition counter is when a transition occurs in the middle of bit sequence, where the outer of the 3 bits are in different states and the centre state is equal to one of its adjacent bit states. This can occur in the bit patterns 001, 011, 100 and 110.

It can be seen from the "A" and "B" sequences in Figure 63 that if the centre bit (b_i) changes state, then the transition counter still counts the same number of edge changes and the error goes undetected. Similarly from the "C" and "D" sequences it can be seen that a single-bit error in the middle of the patterns 100 or 110 will also be undetectable by a transition counter. There are $(m - 2) \times 2^{m-1}$ such possible midstream errors in an m-bit sequence.

Adding these two sources of undetectable errors together, we arrive at a total of $(m - 1) \times 2^{m-1}$ sequences that contain a single bit error that cannot be detected by transition counting. The probability of failing to detect a single-bit error is thus:

$$P = \frac{(m - 1)2^{m-1}}{m \times 2^m} = \frac{(m - 1)}{2m}$$

which for large m is approximately 50 per cent.

In practice, the length of the counter used to hold the transition count will be n bits, or less than m. Overflow is allowed to occur and

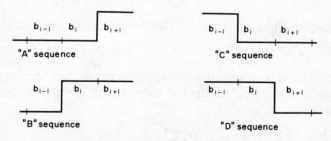

FIGURE 63   UNDETECTABLE MIDSTREAM BIT ERRORS

only the residue left in the register at the termination of the count is used as a measure of the node correctness. This overflow introduces some ambiguity into the measurement and in general for an n-bit transition counter the largest number of detectable errors is $1/2^n$. This overflow gives rise to error rates which are higher than those predicted and in general, more than 50% of all single bit errors will go undetected.

It is clear from the foregoing that transition counting as an error checking technique does not give a very good account of itself in the most likely error event to occur—the single-bit error. For other than single-bit errors, however, transition counting does give a high percentage detection, particularly where the test sequence length is long. For a long period, transition counting was used in *Automatic Test Equipment* (ATE) as the primary technique for troubleshooting logic networks on a nodal basis.

7.4 CYCLIC REDUNDANCY CHECK CODES

In a computer system where data is to be transferred from, say, a disc unit into the computer memory at high data rates, some means of checking and verifying that no errors have occurred during transfer is required. The simple parity bit check is insufficient when blocks of data are being transferred and only finds use in asynchronous, serial data links where information is transferred one character at a time. *Cyclic Redundancy Check* (CRC) codes were developed to overcome the limitations of the simple parity bit check and to provide single and multi-bit error detection in high-speed, synchronous data links.

If we take an 8-bit binary word, then it can be treated as the coefficients of a polynomial of degree 7. The bit pattern

$$B_7 B_6 B_5 B_4 B_3 B_2 B_1 B_0$$

is a shorthand representation of the binary polynomial:

$$B_7 \cdot 2^7 + B_6 \cdot 2^6 + B_5 \cdot 2^5 + B_4 \cdot 2^4 \\ + B_3 \cdot 2^3 + B_2 \cdot 2^2 + B_1 \cdot 2^1 + B_0 \cdot 2^0.$$

As an example, the binary word "1100 0001" can be represented by:

$$1 \cdot 2^7 + 1 \cdot 2^6 + 0 \cdot 2^5 + 0 \cdot 2^4 + 0 \cdot 2^3 + 0 \cdot 2^2 + 0 \cdot 2^1 + 1 \cdot 2^0.$$

For simplicity, let us generalise the variable in the polynomial to be X rather than specifically 2. Thus the same word, using X as a dummy variable, will be:

$$A(X) = 1 \cdot X^7 + 1 \cdot X^6 + 0 \cdot X^5 \\ + 0 \cdot X^4 + 0 \cdot X^3 + 0 \cdot X^2 + 0 \cdot X^1 + 1 \cdot X^0.$$

This polynomial is reducible to:

$$A(X) = X^7 + X^6 + X^0 = X^7 + X^6 + 1.$$

To understand cyclic redundancy checking we will consider that we have a bit stream given by $B(X)$ and a generator polynomial given by $G(X)$. If we now divide the polynomial $B(X)$ by the generator $G(X)$, the result will in general be a quotient $Q(X)$ and a remainder $R(X)$.

$$B(X) = G(X) \cdot Q(X) + R(X).$$

If we now add $-R(X)$ to both sides of the equation, then:

$$B(X) - R(X) = G(X) \cdot Q(X).$$

In a CRC scheme the bit stream to be sent is divided by a generator polynomial and the remainder or residue is appended onto the transmitted bit stream. At the receiving end, the incoming bit stream and residue are divided by the same polynomial used at the sending end to produce the residue. The result of the division at the receiving end of a data link should be remainder of zero because $B(X) + R(X)$ is exactly divisible by $G(X)$, If no residue results at the receive end, then no detectable errors in transmission have occurred.

One of the commonest CRC codes in use uses a 16-bit polynomial given by:

$$\text{CRC-16} = X^{16} + X^{15} + X^2 + 1.$$

To implement this, a 16-bit shift register is used whose inputs are the bit pattern to be checked ($B(X)$), and feedback signals are taken from tappings along the shift register. All the feedback signals and the input signal ($B(X)$) are modulo-2 added (the Exclusive OR operation) which produces a linear sequential circuit because the modulo-2 adder gives the same weight to each input bit. The polynomial is referred to as the characteristic polynomial of the generator, and the actual feedback tapping points are found from its reverse. The reverse equation is found by subtracting each term in the characteristic polynomial from X^{16} in the case of the CRC-16 code, which gives a feedback equation of $X^{16} + X^{14} + X^1 + 1$. Thus into the Exclusive OR gate are fed the input signal and tappings from the first stage, the 4th stage and from the 16th stage of the shift register.

The shift register is synchronously clocked and the input data stream is divided by the characteristic polynomial $X^{16} + X^{15} + X^2 + 1$. At the end of the bit stream, the remaining states of the 16 flip/flops contain the residue, which is clocked out from the register and appended onto the transmitted data stream. Earlier it was shown that the residue is subtracted from the bit stream ($B(X) - R(X)$), but

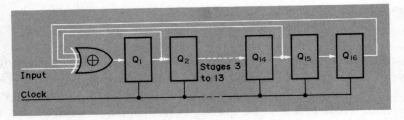

FIGURE 64   LINEAR FEEDBACK SHIFT REGISTER IMPLEMENTING CRC-16

in modulo-2 arithmetic the addition and subtraction operations produce the same results and sending $B(X) + R(X)$ gives the same effect

The CRC-16 generator polynomial has an even number of terms and is reducible to CRC-16 $= (X + 1)(X^{15} + X + 1)$. The $(X + 1)$ factor is deliberately included to ensure that all undetectable errors have even parity. An even number of error bits has to occur to cancel an error, which tends to cluster them. A cyclic redundancy check scheme, however, will always detect single-bit errors in its input stream. CRC codes provides a method of detecting all single-bit errors in a data stream and a high percentage of all multi-bit errors. Hence they find considerable use in high-speed digital transfer schemes involving blocks of data rather than single characters.

Both transition counting and CRC codes are techniques for the detection of errors in long data bit streams. The former, while having a high probability of success for multi-bit errors, is poor at detecting single-bit errors, while the latter will always detect single-bit errors but also provides reasonable performance for multi-bit errors. We would ideally like a testing method which could be applied in the same way that transition counting is applied to single node testing in a logic network, but have the error detection performance of CRC codes or better. Signature analysis, developed by Hewlett Packard, provides this marrying of the two techniques.

7.5 SIGNATURE ANALYSIS

In a CRC scheme, when an input bit pattern is applied to a linear sequential circuit, which implements the division of the bit stream by some characteristic polynomial, we are left with a residue in the shift register. usually this residue is appended onto the transmitted bit stream as an error detecting code. If, in place of adding it onto the bit stream, we display it as a residual count, then the value will be unique for the input bit pattern. By using storage elements, the circuit influenced by all past and current events, and can hence be used on very long data streams. The uniqueness of the residue for a particular input acts like a fingerprint for that pattern and can be used to identify it. The dependency of the residue on the input bit stream has led to the name "signature" being coined for the register residue. By taking the concept of a "time window", using a start and stop pulse and a clock from the system under test, a repeatable pattern can be applied to a node in a logic network. This pattern becomes the input bit stream for a linear sequential circuit, and at the reception of the stop signal the shift register will contain a "signature" for that node and that particular test pattern. This is the concept of signature analysis. The actual value of the signature is irrelevant, but must be the same for the same node when stimulated by the same test pattern and using the same start, stop, and clock signals.

To obtain the best results from the shift register technique, a maximal length sequence is required. This forms the basis for a range of circuits called *Pseudo-random Sequence Generators*. If a 16-stage shift register is used, there are 2048 possible ways of arranging the feedback tappings to satisfy this criterion. CRC-16 uses an even number of inputs, which tends to cluster its errors, while for nodal testing we would prefer a method which spread the errors as much as possible. For the same reason, the tappings should not be arranged to be 4 or 8 stages apart, because these correspond to the bit sizes of the microprocessors most likely to be tested. Hewlett Packard chose an uneven number of inputs using the irreducible feedback equation $X^{16} + X^{12} + X^9 + X^7 + 1$. This corresponds to the characteristic polynomial $X^{16} + X^9 + X^7 + X^4 + 1$. At this point we have to remember that we are trying to develop a general purpose test instrument which can be applied widely to testing digital systems; there are other possible characteristic equations which satisfy the criterion, but this is the one that has been chosen.

In essence, we have a portable instrument which consists of a 16-stage shift register with a modulo-2 adder at its input, which uses start and stop signals to gate the input signal into the circuit and an SUT derived clock signal to shift the data through the register. At the end of a counting period, the information left in the register is taken out and displayed as a characteristic signature for the node under test.

Figure 65 depicts the basis for a signature analyser in which a 16-stage shift register is used. The input bit stream from a node under test is gated, using start and stop signals, into the register which is clocked from a signal derived from the SUT. The residue left in the register after the bit stream has elapsed is taken out and displayed on

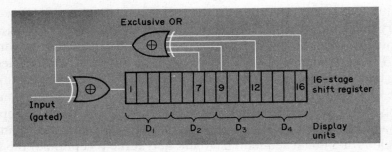

FIGURE 65   THE PSEUDO-RANDOM BINARY SEQUENCE GENERATOR USED IN A SIGNA-
TURE ANALYSER

suitable indicators as 4 hexadecimal characters. This 4-character display is the "signature" for the tested node.

The normal hexadecimal character set was replaced by Hewlett Packard to eliminate any confusion that might arise between the numeric value "6" and the lower case alphabetic character "b". To prevent any ambiguity or confusion the chosen display set was:

BINARY CODE	DISPLAY	NORMAL HEX
0000	0	0
0001	1	1
0010	2	2
0011	3	3
0100	4	4
0101	5	5
0110	6	6
0111	7	7
1000	8	8
1001	9	9
1010	A	A
1011	C	B
1100	F	C
1101	H	D
1110	P	E
1111	U	F

The majority of manufacturers who produce signature analysers have adhered to the Hewlett Packard coding for displayed data. There is no such thing as an almost right signature; the display code 006A is unrelated to the code 006C. Either the signature is exactly right or it is wrong.

7.6 THE PROBABILITY OF SUCCESS USING SIGNATURE ANALYSIS

Before attempting to calculate the probability of detecting an error successfully using the signature analysis technique, we will examine the operation of a smaller linear sequential circuit. The circuit will consist of a 4-stage shift register and a modulo-2 adder. This type of circuit is often referred to as a *Maximal length* or *Pseudo-random Sequence Generator* (PRSG).

A shift register may be described using a delay operator D, such that $X(t) = DX(t - 1)$. Multiplying by D is equivalent to delaying data by one unit of time. We are only concerned with synchronous systems, so that events will always occur at regular intervals. In a computer based system this period will be defined by the system clock.

For the circuit shown in Figure 66, the data entering the register is the modulo-2 sum of samples taken after one clock period and four clock periods along with the input data itself. The feedback equation may be written as $D^4X(t) + DX(t) + X(t)$ or simply as $X^4 + X + 1$. Feeding a data stream into a PRSG is equivalent to dividing the data stream by the characteristic polynomial of the generator, which for this circuit is given as $X^4 + X^3 + 1$. We can demonstrate this by taking an assumed input of, say, 1 000 000 000 000 000 and carrying out the division by a longhand method. We can then check that the result is correct by describing the stage sequence as the data stream is clocked through the register. If the assumption is true, then both methods will produce the same result.

We will start with the PRSG having all its stages set to zero. (This is an invalid state for a normal PRSG which can cycle through $2^N - 1$ states and cannot have all its outputs at zero.) If we consider the circuit in Figure 66 without an external input line, then if all the stages are at zero then the output of the Exclusive OR gate will be zero and when clocked the register will simply shift a zero through itself. The circuit is latched at this state and cannot exit from it. When used as a

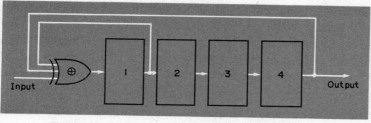

FIGURE 66   A 4-STAGE LINEAR SEQUENTIAL CIRCUIT

true pseudo-random sequence generator, some means of setting at least one stage initially to a logic "1" has to be incorporated to allow the PRSG to be self-starting. In our case, however, the initial all zeroes state is allowable, because the internal sequences of the register will be perturbed by the input sequence. If the input sequence is logic "0"s, then the register will always remain with all its internal states at zero.

CYCLE	$Q_4Q_3Q_2Q_1$	NEXT INPUT TO REGISTER	NEXT INPUT
0	0 0 0 0	1	1
1	0 0 0 1	1	0
2	0 0 1 1	1	0
3	0 1 1 1	1	0
4	1 1 1 1	0	0
5	1 1 1 0	1	0
6	1 1 0 1	0	0
7	1 0 1 0	1	0
8	0 1 0 1	1	0
9	1 0 1 1	0	0
10	0 1 1 0	0	0
11	1 1 0 0	1	0
12	1 0 0 1	0	0
13	0 0 1 0	0	0
14	0 1 0 0	0	0
15	1 0 0 0	1	0

The register is clocked through the 16 input bits and assumed to have been set to zero initially. This is represented by the first line in the above table as cycle 0, in which all 4 flip/flop outputs, Q_4 to Q_1 are at logic "0", and the first bit of the data stream (logic "1") is the next input. The modulo-2 adder at the input sums the input bit with the states of Q_4 and Q_1. If the addition is an odd number, then the output of the adder will be a logic "1". If the addition is an even number, the output will be a logic "0". For cycle 0, the addition of the three inputs is an odd number which will set the adder output to a logic "1" and is shown in the table as the next input to register.

The bit stream from Q_4 is 0 000 111 101 011 001, which is found by reading the Q_4 column downwards, starting from the top of the list.

The characteristic polynomial for the register is $X^4 + X^3 + 1$ which fully expanded is $1 \cdot X^4 + 1 \cdot X^3 + 0 \cdot X^2 + 0 \cdot X^1 + 1 \cdot X^0$. The binary number represented by this equation is 11 001. If we take the input data stream and divide it by 11 001, then the operation proceeds as follows, remembering that modulo-2 addition and subtraction are the same and there are no carries:

```
                                 0 000 111 101 011 001
                        11 001  )1 000 000 000 000 000
                        ↗ ↑↑↑↑   1 100 1
        input             ││││   ─────────
        bit.            Q₄Q₃Q₂Q₁   100 10
                                   110 01
                                 ─────────
                                   10 110
                                   11 001
                                 ─────────
                                    1 111 0
                                    1 100 1
                                  ─────────
                                     11 100
                                     11 001
                                   ─────────
                                      101 00
                                      110 01
                                    ─────────
                                       11 010
                                       11 001
                                     ─────────
                                        11 000
                                        11 001
                                      ─────────
                                            1
```

The result is 0 000 111 101 011 001, which is precisely the same as that obtained by following through the bit sequences in the first method.

The exclusive OR feedback applied to the shift register gives equal weighting to each input bit applied to it. A non-linear circuit will contain combinatorial devices such as AND gates, which are not modulo-2 operators and which will cancel some inputs based upon prior bits.

A linear polynomial is one for which $P(X + Y) = P(X) + P(Y)$. In the example overleaf, three different bit streams, X, Y and $X + Y$ are fed into the same PRSG.

The output from the PRSG is given as Q (quotient) and the residue or remainder as R for each input. Notice that the output sequences follow the Exclusive OR relationship, i.e. $Q(X + Y) = Q(X) \oplus Q(Y)$. Y represents a single impulse bit delayed in time from the start of the main sequence X and the only difference between X and $(X + Y)$ is the single bit due to Y. After the first seven bits, the output bit streams due to X and $(X + Y)$ bear no apparent relationship to one another and using only a 20-bit input sequence, the residues $R(X)$ and $R(X + Y)$ also bear little or no resemblance to one another: $R(X) = 0111$ and $R(X + Y) = 0100$.

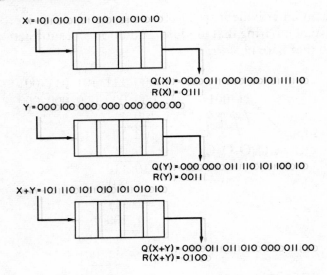

If we consider X as a valid input sequence and Y as an error bit, then the resulting input to the PRSG will be $(X + Y)$ because the theorem of superposition applies to a linear circuit. Clearly from the above, even a single-bit error is detectable as a change in the correct output sequence and further, the output sequence remains erroneous, regardless of the length of the input sequence. The residue left in the register after a finite number of input pulses remains different from the residue due to the correct sequence. In the example, $R(X + Y)$ differs from the correct residue $R(X)$ and the effect of the error remains, even though the error has disappeared many clock periods ago.

The residue depends upon every bit that enters the register and is a characteristic of the PRSG and the input data stream; the residue is hence referred to as a SIGNATURE.

7.6.1 Error detection using signature analysis

A PRSG can be used to detect a single bit error in a data stream of any length.

Assume X is a data stream of m bits, P is an n-bit PRSG and P^{-1} its inverse, such that $P \cdot P^{-1} = 1$. Q is a quotient and R a remainder.

$$P(X) = Q(X) \cdot 2^n + R(X). \tag{1}$$

Take another m-bit sequence that is not the same as X and must therefore differ by another m-bit error sequence E such that:

$$Y = X + E.$$

Now,

$$P(Y) = Q(Y) \cdot 2^n + R(Y).$$

So

$$P(X + E) = Q(X + E) \cdot 2^n + R(X + E).$$

But all operators are linear, so:

$$P(X) + P(E) = Q(X) \cdot 2^n + Q(E) \cdot 2^n + R(X) + R(E).$$

Subtracting (or adding modulo-2) with equation (1) gives:

$$P(E) = Q(E) \cdot 2^n + R(E). \qquad (2)$$

However, if Y is to contain undetectable errors,

$$R(Y) = R(X).$$

It follows that:

$$R(Y) = R(X + E) = R(X) + R(E) = R(X)$$

from which $R(E) = 0$. Substituting into equation (2),

$$P(E) = Q(E) \cdot 2^n$$

and all undetectable errors are found by:

$$E = P^{-1}Q(E) \cdot 2^n. \qquad (3)$$

For a single-bit error,

$$E = D^a(1)$$

where D is the delay operator, a is the period of the delay and 1 is the impulse sequence $100000\cdots\cdots$. Substituting into (3),

$$D^a(1)P^{-1}Q(D^a(1)) \cdot 2^n$$

D commutes with other linear operators, so that:

$$D^a(1) = D^a P^{-1}Q(1) \cdot 2^n$$
$$1 = P^{-1}Q(1) \cdot 2^n$$
$$P(1) = Q(1) \cdot 2^n$$

But by the original assumptions,

$$P(1) = Q(1) \cdot 2^n + R(1)$$

from which, by subtraction:

$$R(1) = 0.$$

However, it has been shown in the previous example that the residue due to a single-bit error is not equal to zero. Therefore, $E \neq D^a(1)$ and the set of undetectable errors E does not include any single-bit errors. In other words, ALL single-bit errors are detectable using a PRSG.

Since X, Y and E are all m-bit sequences, it follows that $Q \cdot 2^n$ must also be an m-bit sequence containing n final zeroes. Q therefore contains $(m - n)$ bits and hence there must be 2^{m-n} sequences that map into the same residue as the correct sequence. Of these, $2^{m-n} - 1$ are error sequences that leave the same residue as the correct sequence. 2^m sequences can be generated using m-bits and only one of these is correct. The probability of failing to detect an error using a PRSG is:

$$\text{Prob(PRSG,fail)} = \frac{\text{Undetectable errors}}{\text{Total errors}}.$$

Hence,

$$\text{Prob(PRSG,fail)} = \frac{2^{m-n} - 1}{2^m - 1}.$$

For long sequences (large m), the probability tends to:

$$\text{Prob(PRSG, fail)} = 1/2^n.$$

A feedback shift register of length n bits will detect all errors in data streams of n or fewer bits because the entire sequence will remain in the register and $R(X) = P(X)$.

For data streams of greater than n bits, the probability of detecting an error using a PRSG is very near certainty, even for generators of modest length. The errors not detectable are predictable and can be generated by taking all m-bit sequences with n trailing zeroes and acting upon such sequences with the inverse of the n-bit PRSG polynomial P. That is:

$$E = P^{-1}(Q \cdot 2^n).$$

For a 16-bit PRSG, the probability of failing to detect an error is:

$$E = 1/2^{16} = 0.00001526$$

from which the probability of detecting all errors is:

$$E = 1 - 0.00001526 = 0.999985.$$

Therefore all errors can be detected using a 16-bit PRSG with a probability of 99.9985 per cent.

7.7 A SIMPLE SIGNATURE ANALYSER

A signature analyser can be constructed at reasonable cost using standard, commercially available components. The basis is a 16-stage shift register with Exclusive OR feedback. Initially the register is cleared to zero which is normally a "latch up" condition for a PRSG, but from which it can be unlatched by any logic "1" bit in the input data stream. *Start, Stop* and *Clock* signals are derived from the *System Under Test* (SUT) along with the signal from the node being tested. The Start signals is used to enable the Clock signal onto the shift register, so that data can be synchronously shifted through it. The Start signal may also be used to gate the incoming data stream. The Stop signal prevents any further clocking of the shift register and also blocks any further input data. The Stop signal may also be used to latch the residue from the shift register to feed display drivers. The block diagram for a simple logic analyser is given in Figure 67.

The block diagram given in Figure 67 illustrates a number of interesting features. The 16-stage shift register may be implemented

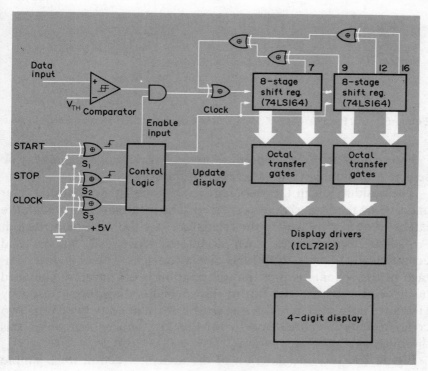

FIGURE 67   A SIMPLE SIGNATURE ANALYSER

using a pair of 74LS164, TTL devices around which Exclusive OR feedback is provided with 74LS86 devices. The input data stream is cleaned up by applying it to a comparator which compares the incoming signal level against a threshold voltage V_{TH}.

In a general purpose instrument, facility has to be allowed for Start, Stop and Clock signals which may be taken to be active on either the rising or the falling edge. A Start signal, for example, may be needed in one test which has a rising edge, but during another the signal may be required to start operations on its trailing edge. Both can be accommodated by using Exclusive OR gates to gate the incoming signals into control logic. In the case of the Start signal, S_1 may be connected to ground or to the positive supply rail, V_{cc}. When it is connected to ground, the output of the Exclusive OR gate is the incoming Start signal, but when S_1 is switched to V_{cc}, the output is the inverse of the incoming signal. If the control logic always demands a rising edge signal to trigger it, then S_1 may be used to select either a rising edge Start signal (S_1 connected to ground), or a falling edge signal (S_1 connected to V_{cc}) to satisfy the control logic criterion.

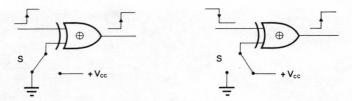

FIGURE 68   AN EXCLUSIVE OR GATE USED TO CHANGE THE SENSE OF AN INPUT SIGNAL

Figure 68 illustrates the point where either edge of a signal may be selected to initiate or stop operation of the control logic.

The residue left in the shift register may be latched into transfer gates when the Stop signal is received. The information stored in the latches is then displayed as the "signature" for the node being tested. A 4-digit display device such as Intersil's ICL7212 can be used economically to drive the 4 LED display devices. The only disadvantage of the scheme is that the information is displayed in standard hexadecimal format and not in the special coding format used by Hewlett Packard. There is a one to one correlation between the two code formats which can be learnt quickly, which overcomes this disadvantage.

Figure 69 gives the diagram of the analyser section for a simple signature analyser. The data input line is buffered using a hysteresis switch configured around an LM311 comparator. The switch has

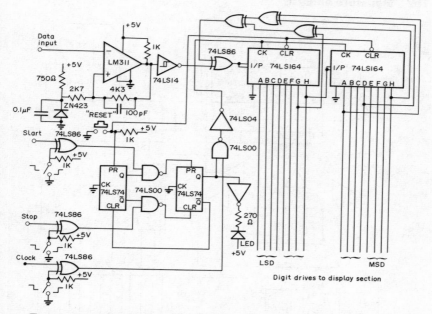

FIGURE 69   A SIMPLE SIGNATURE ANALYSER—ANALYSER SECTION

threshold levels of 0.8 and 2 volts to conform to standard TTL logic levels.

The system is manually reset by the operator prior to taking a signature, which sets the initial state of the control logic and resets the 16-stage shift register outputs to zero. After being reset, the Start signal enables the Clock signal onto the shift register which synchronously clocks data until stopped by the arrival of a Stop signal.

An indication is given that the system is "taking signatures" by an LED connected to the control logic.

A circuit for displaying the signature captured by the analyser section is given in Figure 70.

The display and analyser sections run asynchronously. The outputs from the shift register are buffered through 74LS126, tri-state buffers. Each is sequentially selected from a 2–4 line decoder and its inputs gates onto the output bus and into the ICL7212A display driver. This device will drive up to 4 common anode LED displays in a non-multiplexed mode. The basic clock which controls the display section timing is formed around a CMOS inverter and produces a frequency of approximately 1 kHz. The clock is fed into a 2-bit synchronous counter to provide the four states needed to access each of the buffers.

The data is displayed in standard hexadecimal format with the numeric character 6 having a tail on it to distinguish it from the lower case alphabetic character b.

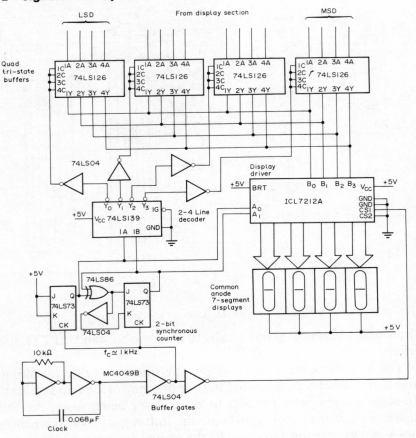

FIGURE 70   A SIMPLE SIGNATURE ANALYSER—DISPLAY SECTION

On some systems it has been found necessary to skew the Start and Stop signals to prevent unstable signal edges from triggering data capture. The circuit given in Figure 71 has been used to overcome this problem.

The circuit shown in Figure 71 is interposed between the outputs of the Exclusive OR gates in the analyser section and the inputs to the control logic. The circuit allows a 100 nanosecond delay between the incoming Start and Stop signals and those used to trigger the analyser. This is sufficient time to ensure that the signals are stable before being used to begin and end data capture.

The signature analyser described in this section is a relatively simple instrument compared to commercially available analysers, but can be used to take signatures from a system and costs approximately 1/10th that of commercial instruments.

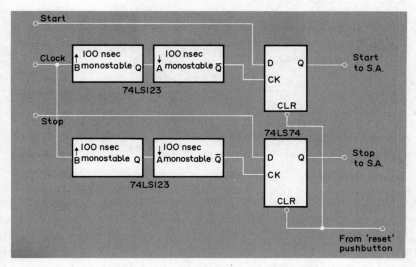

FIGURE 71   DELAY CIRCUIT FOR START AND STOP SIGNALS

7.8 FREE-RUN TESTING USING A SIGNATURE ANALYSER

A signature analyser can be used to verify the kernel of a system by taking signatures from nodes in a free-run mode. This mode of operation is described in Section 2.5, under the heading The System Kernel.

For any test using signature analysis we have to decide which signals from the SUT are to be used as Start, Stop and Clock. On a commercial signature analyser, the lead used to contact any node consists of a logic probe which gives a visual indication of activity. The nature cannot be defined from the probe indicator, but whether or not any activity is taking place can be inferred from its state.

Let us assume that we are dealing with an 8-bit microprocessor system which has a 16-bit address bus. During the free-run cycle, every possible bit pattern will be applied to the address bus, which therefore continues to cycle through a known repeatable pattern. This type of activity is ideal for signature analysis because of its repetitive nature. The most significant address bus line, A_{15}, will be low for half the total number of addresses and high for the second half. One complete cycle of the address bus hence occurs between each successive rising edge of the A_{15} line. We can use this line both as a Start and a Stop signal for the analyser. The first will start and the one following will stop. The residue accumulated in the shift register between these events is then fed onto the display as the signature for

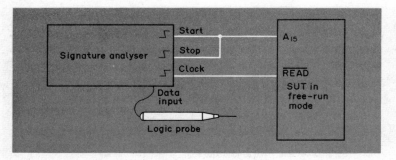

FIGURE 72   TAKING SIGNATURES IN THE FREE-RUN MODE

the node under test. In the free-run mode, all instructions are memory read operations and the $\overline{\text{READ}}$ line can be used as a Clock signal for the analyser. The analyser is set up for rising edges of the Start, Stop and Clock inputs so that data will be synchronously clocked through the shift register on the trailing edge of the $\overline{\text{READ}}$ signal for one complete cycle of the address bus. The arrangement is illustrated in Figure 72.

Before any signatures are taken from nodes in a system, the signature analyser and input signal connections are verified by taking the signature of ground and V_{cc}. The shift register in the analyser is initialised to zero before any data is captured. When the probe is connected to ground, the data input will always be a logic "0" state which will not perturb the initial state of the shift register. At the end of the data capture cycle, the residue left in the shift register will still be zero. Only a logic "1" input can perturb this state, which clearly will not occur when checking the ground signature. Ground then always gives a signature of 0000, which is referred to as its "characteristic signature". The positive supply, V_{cc}, however, is always seen as a logic "1" state which will perturb the initial state of the shift register. The residue remaining at the end of data capture will depend upon the number of input Clock states between the Start and Stop signals and will differ with the trigger signals. For a particular arrangement of Start, Stop and Clock signals, the V_{cc} signature will remain the same and is sometimes referred to as a "characteristic signature" for those particular input connections. It will, however, be different for different sets of control inputs and/or different edge triggering settings. If a node is probed which gives the same signature as V_{cc}, it may be that the node is, through a fault, tied to the positive supply rail. Sometimes, however, nodes which are fault-free do give the same signature as the V_{cc} characteristic signature. The simplest way to differentiate between these two conditions is to note that in the fault-free state the indicator on the logic probe will blink to indicate system activity. If the indicator does not blink, then a fault can usually be inferred.

Any system that uses signature analysis as a test method has to be accompanied by a document which lists the signatures for each and every node in the system. In the free-run mode, although the total address space is cycled through, not all devices in a system will be enabled, because only memory read operations are carried out. Thus an input port, for example, will not be enabled, and no meaningful signatures can be taken from it. In a similar way, the data bus is disconnected from the processor and with the exception of certain conditions will not give meaningful signatures. The list of signatures is taken from a system that is known to function correctly and is documented. At the head of a page of signatures the connections from the system to the Start, Stop and Clock inputs are noted, along with their trigger edge conditions, i.e. whether rising or falling. Also at the head of the document should be given the characteristic signatures for V_{cc} for the stated input connections. This information is used to set up the analyser and to verify the connections by taking both ground and V_{cc} signatures. Signatures are then taken from nodes and the results compared to the documented signatures. An example of a table of signatures for the free-run mode is shown overleaf.

The table is only meant to represent a small portion of the complete list of signatures which will be taken for every device in the system. At the start of the table are given the switch settings and the system signals which have to be used.

The characteristic V_{cc} signature is given as 0001 in this example, which will be checked before taking any signatures from the system. In the free-run mode, the data bus signals are meaningless, and this state is indicated in the table as a series of "don't care" states (X). The address bus is, however, exercised and the signature for each address bus line is given.

Following on from the bus states, there are a series of pin out diagrams for every integrated circuit in the system with the signature of each pin written alongside it. Ground always has the characteristic signature of 0000 and is shown on the diagrams as simply GND. To indicate that a signature of 0000 is valid for a pin and is meant to be other than ground, the signature is followed by a B to show that the LED located in the logic probe tip will blink when a signature is taken. Pin 18 of IC2 is an example of such a condition. Particularly in the free-run mode, many IC pin signatures will have no meaning and will be indicated on the diagrams as an X. An example is pin 3 of IC4. Another condition frequently met is indicated by pin 1 of IC2. Here the signature is 0000, but the probe tip does not blink. Pin 1 for this

Freerun address signatures.

S.A. Switches	Connections
START ⌐	A_{15}
STOP ⌐	A_{15}
CLOCK ⌐	READ

V_{CC} SIGNATURE 0001

Data Bus	Address Bus	
D_0 X	A_0 UUUU	A_8 HC89
D_1 X	A_1 5555	A_9 2H70
D_2 X	A_2 CCCC	A_{10} HPPO
D_3 X	A_3 7F7F	A_{11} 1293
D_4 X	A_4 5H21	A_{12} HAP7
D_5 X	A_5 OAFA	A_{13} 3C96
D_6 X	A_6 UPFH	A_{14} 3827
D_7 X	A_7 52F8	A_{15} 755P

IC1

Left pin	#	#	Right pin
GND	1	20	V_{CC}
1293	2	19	0000
1293	3	18	755P
HPPO	4	17	755P
HPPO	5	16	3827
2H70	6	15	3827
2H70	7	14	3C96
HC89	8	13	3C96
HC89	9	12	HAP7
GND	10	11	HAP7

IC2

Left pin	#	#	Right pin
0000	1	20	V_{CC}
UUUU	2	19	52F8
0001-B	3	18	0000-B
0001-B	4	17	0001-B
5555	5	16	UPFH
CCCC	6	15	OAFA
0001-B	7	14	0001-B
0001-B	8	13	0001-B
7F7F	9	12	5H21
GND	10	11	0001-B

IC3

Left pin	#	#	Right pin
V_{CC}	1	20	V_{CC}
0001	2	19	0001
X	3	18	X
X	4	17	X
0001	5	16	0001
0001	6	15	0001
X	7	14	X
X	8	13	X
0001	9	12	0001
GND	10	11	1920

IC4

Left pin	#	#	Right pin
V_{CC}	1	20	V_{CC}
0001	2	19	0001
X	3	18	X
X	4	17	X
0001	5	16	0001
0001	6	15	0001
X	7	14	X
X	8	13	X
0001	9	12	0001
GND	10	11	597C

test is always at a logic "0" level, which gives the same signature as ground, although putting the residue down as 0000 on the diagram is meant to indicate that the pin is not shorted to ground. If this is so, then the legend GND is placed alongside it.

Fault location using signature analysis is a matter of deciding the possibly faulty area and probing it until an incorrect signature is located. Using the circuit diagram for the system and the tables of signatures, the fault is traced back until a correct signature is found, which isolates the fault to the area between the last incorrect signature and the first good one taken.

7.9 ROM TESTING DURING FREE-RUN

While the primary function of free-run testing is to exercise the system kernel and verify its operation, it is possible also to check any ROMs. During free-run operation, the address bus is repeatedly cycled through every possible bit pattern. The CPU is forced by the No-OPeration type instruction jammed into it to carry out memory read operations for every address. A ROM only contains fixed instructions which, in the free-run mode, are all sequentially placed onto the data bus. By using the ROM chip enable line for Start and Stop signals and again using the $\overline{\text{READ}}$ control line as the Clock signal for a signature analyser, the contents of any system ROM can be tested. The analyser will thus only capture data that pertains to the ROM under test, even though the processor cycles through its total address space. Figure 73 represents the situation where one ROM of several contained in a system is checked using signature analysis during a free-run exercise.

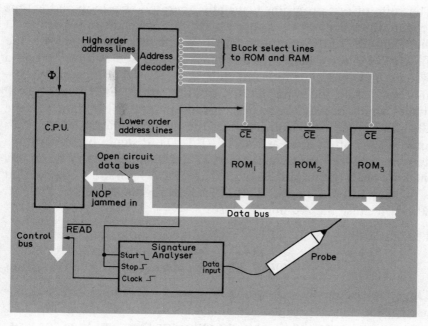

FIGURE 73   ROM TESTING DURING THE FREE-RUN MODE

The same type of test cannot be applied to RAM devices because their contents are indeterminate as a whole and other tests are applied to verify their operation. Similarly, because the CPU only carries out memory read operations, I/O cannot be tested, particularly in an I/O mapped I/O scheme. Input ports in a memory mapped I/O system can be checked during free-run because the CPU treats them in exactly the same manner as a read from a memory location. The inputs to the input port have to be controlled, typically by connecting a test box to it, so that known bit patterns are applied. Output ports in any mapping scheme require a write operation which is clearly beyond the scope of free-run testing.

7.10 SIGNATURE ANALYSIS TEST LOOPS

To test those parts of a system that are not exercised during the free-run mode of operation, special programs have to be written and run. Each is intended to test one part of a system, say an input port, and is usually only a few lines of assembler code long. These signature analysis test programs are usually contained in a ROM, located within the system but not normally used. The ROM may be brought in by either switching the $\overline{CE}$ line from the first system ROM and then resetting the system, or by forcing a RESTART type instruction into the processor and locating the test ROM at the restart address. The suite of test programs is usually operated as a loop and is thus repeatedly executed when the test ROM is brought into circuit. At the beginning of the complete test loop, the line used to provide the START signal to the signature analyser is pulsed and at the end a system line is designated and pulsed as the STOP signal for the analyser. often the same line is used for both START and STOP signals which precludes the need for a separate STOP signal. In the 8080-Z80 type of microprocessors, the most significant address bus line, A_{15}, is often used as the START/STOP signal by executing a dummy input or output instruction. These microprocessors use only the lower half of the address bus to carry an I/O port address, but that is duplicated on the upper half. Thus reading or writing to I/O port 80_{hex} will cause the A_{15} line to pulse.

```
TESTLOOP:        DI              ;disable interrupts
                 IN  80H         ;pulse start-stop line for reads
                 OUT 80H         ; pulse start-stop line for writes
```

This simple sequence of instruction, given in 8080 mnemonics, can be used at the start of a signature analysis series of test programs. The instructions cause the A_{15} line to pulse for read and then for write operations. Depending upon the test being verified, the Clock input to the signature analyser will be taken either from the $\overline{\text{READ}}$ or the $\overline{\text{WRITE}}$ control bus lines.

A simple output port test may be as follows:

```
OUTEST:        XRA  A        ;clear the A register
               STC          ;set carry bit to '1'
LOOP:          RAL          ;move set bit 1 place to left
               OUT  (04)    ;send bit pattern to output port
               JNC LOOP:    ;jump back if not done
```

The concept of the test is to move a logic "1" state through all 8 I/O lines of the output port, one bit at a time. Thus 8 writes to the output port will be carried out before the next sequential test is stepped onto. Using A_{15} for the START and STOP signals, and the $\overline{\text{WRITE}}$ line as the Clock input, each output line is probed and the signature found compared against the documented value. A similar test can be written to test each input line of an input port, but they must be put into a known state first. This is most readily achieved by connecting a test box to the input lines to set the states.

7.11 RETROFITTING SIGNATURE ANALYSIS

Ideally a computer system should be considered as a vehicle for signature analysis testing as its design stage. During the system development, provision can be made for free-running and means incorporated to select test loops from a ROM. Where control signals for an analyser have to be produced by gating several signals together, the necessary hardware can be designed in. Once the system has been fully developed, complete tables of signatures for the various tests can be taken and documented.

Many systems have been designed and developed without any form of testing being incorporated, and these can be retrofitted for signature analysis. The first provision must be to provide the hardware necessary to free-run the system kernel and thus be able to check out the vital parts of the computer. Fortunately in most microcomputer systems the microprocessor is socketed and can be removed. The microprocessor socket provides all the signals and power supplies required by a retrofitted free-run scheme. The simplest approach is to remove the microprocessor and plug into its socket a harness, which is strapped up to open circuit the data bus and to force a NOP type instruction onto the free end of it. The microprocessor is plugged into the opposite end of the harness. For a particular microprocessor, a test department need only carry one harness to be retrofitted into any system that employs that device. Where several different systems have to be maintained and which use different microprocessors, several harnesses can be made up for each device. The power of the free-run mode can thus be implemented into any system whether it was designed with signature analysis in mind or not.

Free-run tests can only exercise part of a system and to stimulate other parts programs have to be written and committed to ROM. The usual approach is to replace the first system ROM with the signature analysis test ROM as a plug-in replacement. The system is then reset and causes the test programs to exercise the complete system.

Signature analysis test programs are run as a series of short items which follow on from one another. At the end of the series the processor is jumped back to the beginning to form a complete test loop. To isolate faults in a system, separate Start and Stop signals are usually needed to test individual components. The majority of these signals are available already in the system; as an example, every ROM may be tested using its own chip select signal from the address decoder to provide Start and Stop control lines. RAM is usually tested by executing an initial series of write operations to it which cause a known bit pattern to be stored in every possible RAM location. These write tests are used to verify the address bus lines to the RAM and the control bus lines that enable the device. Having established that these lines are functional, a series of read tests can then be carried out which verify the data bus connections from the RAM under test. Any signals that are used as Start and Stop signals to a signature analyser should themselves be testable through a previously run program. Ideally, all tests should build on previous ones which verify parts of the system. Free-running a system allows the kernel to be tested, which if functional can be used to provide signals which have been verified. In most cases a satisfactory free-run check is taken as an indication that the microprocessor itself is functional.

7.12 LIMITATIONS OF SIGNATURE ANALYSIS

Only events synchronised to the clock signal used to obtain signatures from a system can be captured by a signature analyser. A microprocessor is a synchronous, sequential machine and the majority of its parts may be tested using signature analysis. However, certain events may occur asynchronously to the system clock and these cannot be checked using the technique. An RS232C serial data link may be operated using a clock that runs asynchronously to the system clock and this will preclude signature analysis as a test method.

It is possible in this instance to take signatures from the serial line using the UART clock as the basis of the analyser clock input. The computer has, however, to be programmed to send a repetitive signal to the line to ensure that a known, repeatable bit sequence is provided for the test.

A bus-structured computer may be viewed as a closed loop feedback system in which a data request is made over the address bus and the response is then fed back over the data bus. The problem with this structure is that a fault in the loop propagates around it, causing apparent bad signatures at locations where no fault exists. Signature analysis does not indicate where within the acquisition time window a fault first appears, which in itself can be used to isolate the faulty component. A bad signature at the end of the time window gives no trace history. Thus when many devices output onto a common bus and a faulty signature occurs, signature analysis provides no clues as to which device is causing the problem.

Trace analysis is an enhanced form of signature analysis which is used to locate a fault within the Start-Stop window. A trace analyser is programmed (usually manually) with a sequence of intermediate signatures and their corresponding time values within the window. When a trace is taken, the first failure of the measured data when compared against the stored information causes the analyser to halt and display the time value of the error. The fault can then be traced back to which device was operative at the time of failure. Valid trace analysis data, in common with signatures, has to be taken from a known working system and documented for it. This involves separating signatures into shorter sections with a consequent increase in the amount of documented information. This extra documentation is fortunately only required for those parts of a system that cannot be tested adequately using conventional signature analysis.

7.13 SIGNATURE ANALYSIS AS A GENERAL TEST TECHNIQUE

Signature analysis is a simple test technique, consisting of probing individual nodes in a network and comparing their signatures with documented values. Intelligent application of the method can be used to identify quickly a faulty area and from there to isolate the faulty component.

Good documentation is essential if signature analysis is to be applied as a primary test technique. Any modifications entail retaking every signature for a system, and updated tables must be supplied whenever modifications or upgraded schemes are developed. The importance of maintaining up-to-date documentation cannot be over-emphasised. In many instances systems are modified, but their circuit diagrams are not kept up to date. This in itself can lead to testing problems, but if the signature tables are not maintained, then the technique will be invalidated and unworkable because the signatures taken will be meaningless without their corresponding, correct documented values.

Chapter 8

Emulation

Emulation is the process by which one system is used to copy the actions of another. There are several levels at which a system being developed can be emulated. An operating environment can be created in one computer which attempts to simulate the actions to be expected in another, independent system. Thus one system is made to behave as if it were another and the actions of the simulated system may be studied from the behaviour of the simulating system. We create, in effect, an unreal environment which will, in general, not behave in precisely the same manner as the system it attempts to simulate. Usually the simulating system will operate at a slower rate than the simulated system because we are attempting to duplicate the actions of hardware through software. The simulation is only a model of the system being developed and will lack certain features of it, such as operating speed and real inputs and outputs. Simulation can be thought of as the lowest level of emulation.

At the other extreme, we can build the final system and use diagnostic aids such as a logic analyser to extract information from it to test and prove its integrity. At this prototype stage in development, there are likely to be many possible sources of error from faults in the hardware and software bugs. The system may then be afflicted with many faults, which in isolation may be simple to solve, but which taken together may prove very difficult to diagnose.

Practically, we would prefer some system which allowed us to cover the middle ground between these two extremes. Simulation can provide a vehicle for software development, but we would also like to extend the hardware features of the developed system in gentle steps. For example, we may wish to use the emulating system to provide a clock source for the one being developed and having proved that part of the circuitry allow the new system to operate on its own clock. The emulating system can then be used to check the target system's operation by using it as a diagnostic aid. Further, we would like to develop the final system's software in the emulating system's environment and transfer it into the target machine a block at a time. Thus as each part of a program is developed, it can be tested in its final environment, a piece at a time, until the complete program has been loaded and checked.

The ability to control the actions of a target machine from the operating environment of another has been called *In-circuit Emulation* (ICE) by Intel, who originally developed it as a working development tool. The machine used to provide these facilities is termed a *Development System.*

Development systems offer a wide range of facilities when developing a prototype, and also provide diagnostic features for troubleshooting. Instruments that provide some of the features found in a development system for diagnosis of hardware faults are referred to as free standing emulators.

8.1 DEVELOPMENT SYSTEMS

A development system is a microcomputer which provides facilities for both software and hardware development. Software is catered for by a suite of programs which collectively is called an *Operating System* (OS). The size of the operating system means that it has to be kept on a floppy disc (or hard disc) and segments of it are loaded into the computer as and when required. Some of the programs are so large that they have to be loaded into the main computer memory a piece at a time. Disc resident operating systems are referred to as *Disc Operating Systems* (DOS). A typical DOS will contain the following, non-exhaustive list of programs:

An Editor,
A File Manager,
A Linker/Loader,
An I/O device handler,
An Assembler,
A Compiler,
A Memory Manager,
A Debugger,
A Real Time Analyser,
A PROM Programmer.

In a specific system some of these programs will be called by other names and some may be joined. A *Real Time Analyser* program may, for example, form part of a *Debugger* program.

A development system will also provide hardware features, in excess of those found in a general purpose microcomputer. The extra features, may include:

A Memory Management Unit (MMU),
A Real Time Trace Analyser,
An In-circuit Emulator,
A PROM programmer.

These extra features make a complete development system, an expensive item in comparison with a general purpose microcomputer. The features are, however, necessary if a complete target system is to be developed in the shortest possible time. To spread the extra costs involved in a development system, particularly where several different microprocessors are in use, several general purpose development systems are available which can cater for a number of different devices. The alternatives to these are development systems produced by the microprocessor manufacturers, which only cater for their particular devices.

The operating system provides the link between the user and the hardware of the physical development system. The user can configure the machine to suit his personal requirements by invoking the programs that form the operating system to recognise and allocate I/O channels to peripheral devices and partition memory space between the development system itself and the target machine. Ideally, the operating system should be "user friendly" in the sense that it is easy to understand and use. Many early operating systems were far from this ideal and needed as much study as the microprocessor system being developed. Later systems overcame this distinct disadvantage by providing "softkey" facilities which prompted a user to select a function by labelling on the CRT screen the functions allocated to several keys located just below the prompt characters. On power-up, for example, the unmarked softkeys might be labelled EDIT, LOAD, ASSIGN, DEBUG, etc. If the user wished to enter the Editor program, he would press the key labelled EDIT which in turn would call into main memory the Editor program. Once into the Editor program, the softkeys would be relabelled on the screen with NEW FILE, OLD FILE, QUIT. If the user then wanted to call up a previously written program, he would press the softkey labelled OLD FILE and then be prompted for the file name. Softkeys eliminate the need for a user to have to learn a set of calling codes for operating system programs and provide an ergonomic interface between the system and the user.

8.1.1 The Editor

The *Editor* program allows the user to enter new programs or program segments and to save them on some mass storage medium such as floppy discs. It also allows previously saved programs to be called down from disc and modified. The editor then is the primary medium for entering the software that will eventually run in the target machine.

Programs may be entered in a high-level language such as Basic or Pascal, or in Assembler language. The editor itself makes no distinction between the type of code entered and provides no error-checking facilities. The translation from a high-level language or assembler is left to other programs within the operating system, which also provide the error-checking features.

When a program is saved as a file, an extension is often tagged on to distinguish its type. If, for example, there is a program in Basic which has been saved on disc and has the title ANALYSER, then in some systems it will be saved as ANALYSER·BAS. Similarly, a program written in Assembler mnemonics with the same title may be saved as ANALYSER·ASM. The extensions will be checked by other programs within the operating system when activated, in order to verify that it is attempting to process the correct type of file. There is no point in trying to compile an assembly language program, and if it is attempted, then an error message will be printed on the screen to inform the user that he is attempting an invalid procedure.

When used to modify a program, most editors are either line or screen. A screen editor will display at any one time the maximum number of lines that can be held. Thus, if the display has a total of 24 lines, then 24 lines of the program will be shown. Scroll keys are provided to allow the user to move the displayed portion of the program through the program list in either a forward or backward direction. They permit the user to search rapidly through the program to find the section he wishes to change. Having located the area, the user may then insert or delete code, or modify the values of existing code. In a screen editor, cursor control keys are used to position the cursor on the area to be changed anywhere on the screen. Let us say that the programmer wishes to enter a new line of code between two previously entered lines. By placing the cursor over the second line and involking the INSERT mode, a space will be created between the two lines on the screen by the editor. The programmer may then type in the new line between them to create his new program segment. The two old lines may be positioned anywhere on the screen and the cursor controls used to identify the second line for insert. Cursor movements are also allowed horizontally, so that it may be placed over a character on a previously entered instruction. The character may be deleted or altered or moved to the right to allow an extra one to be inserted.

A line editor places two horizontal lines on the screen and data may only be entered or modified between them. Thus if a new instruction has to be entered in a previously written program between two which already exist, then the second has to be placed between the lines on the display. The scroll keys again permit the program to be rolled up and down the screen so that desired position may be located within the line window. Invoking the INSERT mode will then cause the program from the second instruction to be shifted down one line to create a space within the line window. The new line may then be entered in the empty display window. Where a character is to be edited on a line, first the line has to be placed within the window and then it may be altered using the cursor controls to locate the point on the line from which modifications are to be made.

The choice between a screen editor and a line editor is usually a matter of preference, and many systems cater for both. The user is then left to use the editor that he finds most suitable.

8.1.2 The File Manager

The *File Manager* takes care of the saving and loading operations between the computer's main memory and the mass storage medium employed in the system. It may form an explicit part of an operating systm and will have to be invoked to load into the computer, the editor program say, or it may be implicit in the operating system and be invoked through the use of other commands such as calling for the editor from one of a number of softkeys. In the latter case, the file manager is transparent to the user and is brought into action from an overall supervisory program.

In those systems where the mass storage medium has no direct "intelligence" associated with it, the file manager may also take care of general housekeeping functions. These will include maintaining a block availability map of free space in the mass storage medium and generating link addresses when sectors have to be concatenated to allow for the size of a stored program.

8.1.3 The Linker/Loader

The complete program for a system, either because of its length or because it has been written by several programmers, will have been written in segments. Each segment will have been stored as a separate file on say a floppy disc. The segments may have been written in a high-level language or for the majority of dedicated system programs will have been written in *Assembler* language. When the final program is to be assembled into machine code, the segments will have to be run through the assembler program for conversion. The *Linker* allows the programmer to specify to the assembler program, which files are to be assembled and in what order. By having to define only the order at this stage, the programmers are left free to write each program segment in whatever order they find most suitable.

Programs that are run through an assembler are called *Source* programs and the resulting machine code program is termed an *Object* program which is itself stored on floppy disc as an object file. The second part of the *Linker/Loader* program allows this runnable object file to be loaded into memory starting at some specified location. The machine code program loaded in this way can be executed directly by forcing the CPU program counter to the start address of the program.

8.1.4 The I/O Device Handler

A development system can have a wide range of peripheral devices attached to it. These may vary from a *Visual Display Unit* (VDU) to a high-speed line printer. The mass storage medium associated with the system will also be treated by the machine as another peripheral device. All the devices that can be attached to the development system require handling programs which take the data from the system and convert it into a format suitable for acceptance by the peripheral. Thus format conversions are transparent to the user of the development system with the complexity hidden in the device handling programs. The user has to allocate I/O channels to peripheral devices such as the VDU screen and keyboard and can then send or access information through them. For example, the VDU screen may be designated by the label CONO and allocated to channel 0. Any write to channel 0 will then appear on the VDU screen. Similarly, the keyboard may be labelled CONIN and allocated to channel 1. Any read from channel 1 will cause a keyboard character to be entered into a program.

Usually peripheral devices such as the VDU screen and keyboard have default channel numbers which may be changed if desired. Any other devices may be added and their handling routines maintained as part of the operating system by executing a *System Generation* program at the commencement of operations and linking the device handler into the operating system.

I/O device handlers create a flexible scheme and allow standard peripheral devices to act as simulating I/O for system development.

8.1.5 Assemblers

The lowest level at which the majority of microprocessors can be programmed, is machine code. Machine code binary numbers are the bit patterns that the microprocessor decodes to carry out the instruction implied by the bit pattern. For convenience, hexadecimal values are used when programming because they are more convenient for a human operator to remember. Programming at the machine code level is prone to error and difficult to amend when absolute addressing has been used. Every machine code instruction set has a corresponding set of *Assembler* mnemonics with a one to one correspondence between a particular machine code instruction and an assembler mnemonic. Programming in assembler mnemonics then is as efficient as direct machine code programming, but is less prone to error because of the use of meaningful labels in place of numbers. A program written in assembler mnemonics has to be translated into machine code before it can be run on a computer system; this task is carried out by an assembler program. A program listing in assembler mnemonics will be entered into a development system using the editor and then saved as a file on the mass storage medium. Most assemblers permit four fields, into which information is entered. These fields are often called LABEL, OPCODE, OPERAND and COMMENT fields. The LABEL field allows addresses to be referred to by a name rather than as an absolute address. This means that programs may be written using labels in place of addresses and the actual addresses left until the final assembly operation. Into the OPCODE and OPERAND fields are entered the actual assembler mnemonics for the instructions while the final COMMENT field allows the programmer to sprinkle comments liberally about the program through it and thus, hopefully, make it more readable. If a program is referred to some time after it has been written, a programmer can understand its function from well-written comments. A simple example of an assembler listing might be as follows:

```
LABEL    OPCODE    OPERAND    COMMENT
START    LD A,     06H        ; load reg A with first output
         OUT(08H)             ; number
LOOP     INC A                ; increment output
         OUT(08H)
         JP NZ,    LOOP       ; loop until new output is zero.
```

No addresses have been specified for the program and the programmer has only to remember the mnemonic codes for instructions and not a set of hex numbers as is the case for a machine code program. Note that after numbers, a capital H has been placed to inform the assembler that the numbers are in hexadecimal format. Most assemblers permit the use of decimal, hexadecimal and binary number formats. The comment field (usually delimited by a semi-colon) plays no part in the assembly operation and merely acts as an aid to the programmer.

For every microprocessor, there has to be an associated assembler to convert the specific mnemonics of that processor into its unique machine code values.

No machine can convert the above program into machine code values without extra information about where the program is finally to reside in memory. Such vital information is called an *Assembler Directive* and has to be given at the beginning of the assembly operation. The mnemonic ORG for origin is used almost exclusively by all assemblers as a directive for the start address for the program being assembled. The directive ORG 1000 will cause the assembler to allocate the opcode of the first program instruction to address 1000 hexadecimal. Where jumps occur in the program, the assembler can work out the addresses that have to be put into the machine code program from a knowledge of the instruction set and the initial address specified by the ORG directive.

The assembler has to be informed of where to end program assembly. This is usually given by the END directive following the last program instruction in the source program listing. Other directives are used to reserve storage space in memory for the results of intermediate calculations and/or look up tables. Any variable referred to by a label in the source program has to be declared as a numeric value at the beginning of the assembly operation, again by using an assembler directive. For example, an input port used to read in a set of switches may be referred to as SWIT in the program. To enable the assembler to convert the program into machine code, SWIT has to be defined as a numeric port address. The equate directive EQU is often used for such a task. If, say, the port referred to as SWIT was actually input port 04, then the assembler directive SWIT EQU 04H would be given at the beginning of the assembly operation.

Many assemblers are called *Macro-Assemblers* because they permit macros to be written and included in the machine code object program. A macro is similar to a user defined function in a high-level language, is written once and then referred to by a label. We may have written a macro to multiply two 16-bit numbers together and called it MULT. Within the source program, we can include the assembler directive MACRO MULT which will cause the macro to be inserted at that point. A macro differs from a subroutine in that it is written once but inserted as a block of code whenever it is specified.

Assemblers are widely used to produce machine code programs for systems used in engineering. The assembler mnemonics have a one-to-one correspondence with machine code values and are thus as efficient as a program written directly in machine code. An assembler program is less prone to error when entered because of its improved readability for a human operator.

8.1.6 Compilers

A programmer, when writing a program at the machine code or assembler levels, has to remember a great deal of detail about the variables used in the program and the way in which they are manipulated. Any operation has to be built up from a sequence of simple instructions until the final result is achieved. Where a large program is being written, involving several programmers, there has to be a great deal of communication between them about data value types, where they are stored and how one program manipulates them before being passed on to another segment. Human operators find the English type statements of high-level languages easier to program and understand. Where time is not a constraint, a program may be written in a high-level language such as Basic or Pascal and then run through a program called a *Compiler* which translates it into machine code. The translation process may either be direct or via an intermediary language. The intermediary languages are sometimes called pseudo-codes. One form of the latter process is to compile into assembler mnemonics which are then assembled into machine code.

The translation process is inefficient because the majority of compilers will take a high-level language statement and convert it into a fixed sequence of assembler mnemonics. The coding used is general in the sense that it has to be used every time the same instruction occurs in the high-level language program. Usually, for any specific case, a more efficient listing can be produced directly in assembler mnemonics. Data types are usually fixed, which will generally allocate more storage space for a variable than it requires. Most compilers do not search for repetition in the program which can be removed as one or more subroutines.

Compiled programs will take up more storage space and take longer to execute than equivalent programs written directly in assembler or machine code. In many cases the increased storage and execution time are not serious problems because they are more than compensated for by the relative ease with which the program can be written. Program development time is reduced with a corresponding saving in development costs.

Often only parts of a program are time critical: we may, for example, require a delay routine with a well-defined time. Most compilers allow for such situations by enabling assembler inserts to be included. Thus time critical segments may be written directly in assembler mnemonics and inserted into the high-level language program where necessary.

There are no established rules of thumb to decide when a program is sufficiently large to warrant using a high-level language and compiling it. Many programmers insist on using assembler mnemonics for almost any size program because of the intimate acquaintance it gives them to data values, their whereabouts in memory and the ease with which they can manipulate bits in a word. Assembler is usually manageable up to about 8K, but for any programs in excess of that a high-level language should be seriously considered.

8.1.7 The Memory Manager

The term *Memory Manager* has been applied to several different types of program, one of which dynamically reallocates memory space when programming demands dictate. The memory manager used in a development system, however, is used to partition the system memory between the development system itself and the target machine during simulation and emulation exercises. The development system typically will use an 8-bit microprocessor with a 16-bit address bus, which can directly address any one of 65,536 (64K) memory locations. To cater for some of the very large programs that have to run on the system, such as a compiler, the system will usually be fully populated with memory devices so that every possible address that can be accessed by the CPU will have a corresponding location in memory. When a target system is attached to the development system it will also contain memory which under normal circumstances will conflict with locations within the development system. To overcome this problem, blocks within the development system may be selectively disabled and allocated to the target system. Thus the total memory space is partitioned between the two.

The capacity to disable blocks of memory selectively serves a dual function. At the simulation stage, the target system memory will be disabled and part of the development system memory used to emulate it. Thus simulation programs are run in the development system and the behaviour of the simulated target system is verified from the responses. Clearly this approach is limited because the actual hardware of the target system is not being used, the processor in the development system may be different from the one used in the target system and there will usually be timing differences between them. Simulation exercises are, however, useful as a means of debugging the system software prior to running it in the target system.

At the other end of the scale, after the software has been developed as far as possible in the development system, the programs may be committed to EPROM and plugged into the target system or down loaded into RAM in the target system from memory within the development system. To prevent any conflicts between addresses in both systems, the memory manager may be invoked to disable the address space used by the target system in the development system. The development system is still allowed to access the locations that have been disabled within it and will either write or read from locations within the target system. The target system can now run programs resident in it at its full operating speed and the development system can access information about the target system by interrogating its memory. This situation in itself represents a possible conflict position, because we have two devices capable of accessing common memory.

The development system is often only allowed to access this common memory after certain events have occurred, such as a breakpoint in the program. When this is met, the target system halts and the development system can access its memory to display its state and that of the CPU after the breakpoint.

In many development systems the main system memory and the emulation memory are completely segregated so that the development system can run its programs and also perform emulation operations. Most development systems of this type also contain separate processors to handle functions such as running the memory manager program, with a second dedicated to handling simulation and emulation.

8.1.8 The Debugger

Mistakes in a program are called software bugs; the system program used to locate and alter them is called the *Debugger*. A debugger consists of several elements in a development system; at the simplest level, it carries out the functions of a simple *Monitor* program, while at its highest level, it traces and displays target system operations up to a set breakpoint.

Typical of the monitor level commands "D" for display, which will cause the contents of specified memory locations to be printed onto the VDU screen. Thus if locations 0070 through to 008F are to be viewed, the command form under a typical debugger will be:

$$D \quad 0070 \quad 008F.$$

Memory locations may be changed using the debugger, and the altered program also run under its control.

These types of operation are useful, if after having assembled a program and run it, faults are discovered which can be patched by changing specific memory location values and then rerun. This type of facility is clearly useful at the development and simulation stages.

During emulation the development system is being used to trace the actions of the target system without directly interfering with it. The type of commands needed during this phase of development differ from those needed during software debugging. The debugger will allow a breakpoint to be set and all target system transactions stored when a program is executed up to the breakpoint. Usually only a limited number of transactions can be stored, due to the restricted memory size of the development system. Besides setting breakpoints and tracing program flow, the debugger allows the user to single step a program in the target system, select an emulation mode and program EPROMs.

8.1.9 A real-time analyser

Many development systems contain a *Logic Analyser* which can capture information from a target system in the same way that a stand alone analyser will store information from a system. The analyser will usually contain its own memory, but it is programmed from the development system keyboard and displays its results on the VDU screen. Often because the analyser forms only a part of the complete system, its facilities are restricted compared to a stand alone logic analyser.

8.1.10 An EPROM programmer

During the normal course of system development the main program routines have to be committed to a ROM type device, plugged into the target system and run in their final operating environment. For a large volume product, the final vehicle for these programs will be mask programmed ROMs, but at the development stage EPROMs will be used, in case changes have to be made to the software. A development system should contain facilities to program EPROMs which may be a separate program in the operating system or form part of the debugger suite.

8.1.11 In-circuit Emulators

In order to carry out emulation exercises on a target system from a development system they have to be connected together by an *Emulation* pod. The microprocessor is removed from the target system and placed in the pod which is then plugged into the vacant microprocessor socket. All the signals from the microprocessor are buffered into the development system which can also control it by forcing the states of the microprocessor control bus. Information can be extracted from the target system by monitoring the states of the system busses and storing them in emulation memory within the development system.

One major problem that has to be solved by any emulator is the matching of speed between the target machine and the emulation environment in the development system. The emulator pod itself extends the signal path lengths, which can give rise to timing skew problems between them. As newer, faster devices are developed, matching emulation systems have to be developed to cope with their increased speeds of operation. This applies not only to the micro-processor pod system and its related buffering circuits, but also to the memory used within the development system to store information captured from the target system. For any specific microprocessor device, a matching emulator has to be developed which can extract information and control it in real time. Real time in this sense means at a pace determined by the target system and not some relatively lower speed which suits the development system.

Information is passed between the two systems with the target system operating under pseudo-normal conditions, Its microproces-sor has been removed to an emulator pod, but to all intents and purposes it still appears to have its microprocessor still plugged into itself. Intel coined the name *In-circuit Emulation* (ICE) to cover this type of emulation environment. With increasing clock rates and hence faster information transfers in a system, the design of in-circuit emulation systems is becoming a difficult and time critical area.

8.2 SYSTEM TESTING USING A DEVELOPMENT SYSTEM

The primary objectives of any development system are the develop-ment and debugging of software and hardware during the design and prototyping stages of a product. The same development environment can also be used to troubleshoot a system after it has cleared the development cycle. A development system can have its repertoire extended to include programs which can be used to exercise the target system, via an emulation pod, for fault-finding. The facilities offered by a development system during fault-finding are the use of the debugger programs and the real time analyser. These can be extended to include retrofitted *Signature Analysis* (SA) programs run from the development system. Such extensions can be included by modifying the emulator pod to cater for free-run conditions and writing SA test loops, stored in the development system, which exercise the various parts of the target system. SA test loops usually require START, STOP and CLOCK signals derived from within the target system, which will still entail the use of a separate SA instrument to capture and display signatures.

The majority of personnel involved with testing a system will not have available a development system, and even if there is one it is unlikely to be the one used during the product development. Test programs are hence not readily available to check out a faulty product. This limitation has been partially overcome because, separate, free standing emulators have been developed as low cost alternatives to development systems.

8.3 STAND-ALONE EMULATORS

By removing the socketed microprocessor from a system and placing it in an emulator, the system can be controlled and exercised from the emulator. This concept forms the basis of emulation systems. The control element, the microprocessor, is now in a separate environment which can be programmed by test personnel to carry out exercises on the target system and report any faults in it.

The data bus can be open-circuited within the emulator and a No OP type instruction forced into the microprocessor to put it into a free run mode of operation. The emulator can thus verify the system kernel and check the address bus lines. The opened end of the data bus is also available to the emulator which can thus run free run ROM tests on the target system. A sophisticated emulator does not have to be informed of the memory map of a target system, but can determine the location of ROM and RAM from a knowledge of the processor (memory mapped or I/O mapped) and by carrying out *Checkerboard* tests.

An emulator should be capable of providing several types of RAM test. RAM locations can be tested functionally using the standard types of checkerboard pattern tests. These usually consist of writing to and then reading from every RAM location the worst case bit patterns 0101 0101 or 1010 1010. Many emulators also offer the bit patterns 0000 0000 and 1111 1111 as test patterns for RAM testing.

Some RAM faults occur because of short circuits between address bus lines, either within the device or due to an external fault in the system. Many emulators provide a test which can check for such faults. Initially, all RAM locations are cleared and then one location is written to with a bit pattern, say FF_{hex}. All of the RAM is then read from, to check whether any other location contains the written test pattern. If another location gives the test pattern result, its address is Exclusive ORed, with the address of the location initially loaded with the test pattern to determine the offending address bus lines.

Assuming that all of the RAM has been cleared and address 0800_{hex} is written with FF_{hex}. During the read all RAM program, say, address 0900_{hex} produces the test pattern of FF_{hex}. By Exclusive ORing the addresses 0800_{hex} and 0900_{hex} it can be established that a short exists between address line A_8 and ground.

	A_{15}			A_0
Address 0800	0000	1000	0000	0000
Address 0900	0000	1001	0000	0000
Exclusive OR	0000	0001	0000	0000

       ↖ this bit indicates an address bus
related fault on line A_8.

Similarly, say that address bus lins A_8 and A_9 become shorted together. Writing the test pattern to location 0900_{hex} will cause location 0B00 hex to provide the test pattern when read. A fault can be inferred from the Exclusive OR operation on line A_9.

	A_{15}			A_0
Address 0900	0000	1001	0000	0000
Address 0B00	0000	1011	0000	0000
Exclusive OR	0000	0010	0000	0000

       ↖ this bit indicates an address bus
related fault on line A_9.

The fault can be isolated to either the address bus itself or to within a specific device by qualifying the test with a free run test to check the address bus lines. If the fault appears only during the RAM test, then it lies within the RAM, which will have to be replaced.

Emulators, beside providing fixed tests preprogrammed into them, also cater for user written test programs. These programs may be stored in a plug-in EPROM, or in more sophisticated systems may be written using the emulator as a computing system and stored on cassette tape or floppy disc. The test programs may be signature analysis routines or specific test programs to exercise parts of the system. Emulators thus provide a general purpose vehicle for running retrofitted free-run checks, signature analysis programs and general diagnostic programs to exercise a system. They are generally portable and are consequently far more attractive to field service personnel than a development system.

Chapter 9

Self-test and Diagnostic Software

As equipment increases in complexity, the need for in-built self-checking becomes paramount. An "ideal" system will self-check itself at power on and fail to start if a fault is discovered. It should also indicate the offending part of the system so that it can be quickly replaced and the system down time minimised. In a large system, self-check need only be carried out down to sub-system level, where complete functions are tested on a go/no go basis. Each function should have some simple identifier such as an LED to indicate its status so that first line maintenance is reduced to checking the condition of a series of LEDs to discover the offending sub-system.

While the majority of microcomputer functions can be checked in this way, I/O related systems, particularly when wired into other equipment permanently, cannot be verified easily without exercising the I/O lines. This testing in itself may be particularly hazardous for the equipment connected to the computer. Such systems can be checked, but only by including extra hardware which prevents any unwanted test signals from activating plant based equipment during test. This adds the extra problem of checking the extra hardware, which may itself develop a fault.

Unfortunately most designers address themselves to satisfying the primary purpose of a system and often view testing as a laborious extra exercise. Thus test software and any extra hardware needed to carry out the tests are treated as afterthoughts. The hardware required may involve considerable redesign and is often omitted. The majority of microprocessor based systems currently in use have either very little or no test facilities built into them; a situation which has to be redressed by the test engineers. Test harnesses have to be made which allow a system to be free-run, and diagnostic software has to be written by personnel who have to expend considerable amounts of time learning the system in sufficient depth to permit them to write the test programs. These functions should be incorporated into the system at the design stage by the design personnel who are clearly most familiar with the system.

The problems of retrofitting test facilities have forced many users to demand that these features be designed into the system from its inception. Designers are thus being forced into providing test features in their systems which satisfy customer needs for simple first line maintenance and minimum down time. This utopian situation does not, however, apply to the very large number of microprocessor based products already in use. To cater for their needs, self-test facilities have to be retrofitted and diagnostic test routines written to exercise them.

9.1 SELF-TEST ROUTINES

Self-test routines are considered to be those programs which are initiated during the power-up sequence of a system and which check it in its final operational environment. They thus exclude any tests which have to exercise I/O related functions where potentially hazardous situations can be created by the test patterns themselves. Power on self-test routines usually assume the system kernel is functional or they will not operate. Self-test programs are, in the main, tests on the memory devices of a system and cover both ROM and RAM checks. If these tests execute correctly, they also verify by implication the system address decoding scheme and the data bus. The ROM containing the self-test routines has itself to be functional or at least that part of it which contains the routines. While these constraints may appear at first sight to be very restrictive, if a system fails to power up then the faulty area is at least reduced to a few devices only. The suspect devices can then be tested individually using the free-run mode of operation.

All test programs should be kept short and simple. Rather than attempt to encompass several devices in one, a short separate routine should be written for each one. The basic reason to stick to this simple rule is that each is sufficiently small to guarantee that it will be error-free and can thus be relied upon during the test time. This principle should be adhered to for any test routine, be it a self-test, a diagnostic, or a signature analysis program.

9.1.1 ROM testing

The simplest test to apply to ROM type devices is the *Checksum*. Every location in the ROM is added up and any carries are ignored. The final location in the ROM is omitted from this procedure and is used to store the result of the addition. The resulting sum is then compared with the contents of this final location. If they agree, then the ROM is assumed to be fully functional; if they disagree, then a fault is inferred and a suitable error display activated. A variant on this technique is to store the checksum value in another ROM device rather than in the final location of the tested ROM. For a system containing a large number of ROMs, this approach is to be preferred and overcomes the need to include instructions at the end of every ROM to jump over the checksum location.

A typical ROM test routine will be:

```
START ROMTEST
     START: LD A,00
            LD B,A             ; clear ACC and reg.B
            LD HL, ROMTOP      ; set HL to point to top of ROM
            LD BC, ROMBOT      ; set BC to point to bottom of ROM
ROMLOOP: DEC HL
            LD A,(HL)          ; get number from ROM
            ADD B              ; add previous total in reg.B
            LD B,A             ; transfer new total back to reg.B
            XOR A,A            ; clear carry flag if set
            SBC HL,BC          ; subtract ROMBOT from current value in HL
            JP NZ, ROMLOOP     ; if all locations not added, go back
            LD HL, ROMTOP      ; point HL to CHECKSUM location
            LD A,B             ; bring total from reg B into Acc.
            CP (HL)            ; compare total to CHECKSUM
            JP NZ, ROMFAULT    ; if different, jump to error routine
                               ; otherwise continue with next test.
END ROMTEST
```

The routine has been written in Zilog Z80 Assembler mnemonics and can be applied to any ROM in a system. During the assembly operation directives will be used to equate ROMTOP and ROMBOT to specific address locations. If, for example, a 2K-byte wide EPROM such as the 2716 is being tested and it address space is 0000 to 07FF$_{hex}$ in a system, the two assembler directives

```
        ROMTOP      EQU      07FF
        ROMBOT      EQU      0000
```

will be inserted at the beginning of the assembly operation.

The routine begins by initially clearing the A register (the *Accumulator*) and the B register to zero. The B register is used to store the current running total as each location's contents are added together. The register pair H and L are then pointed to the highest location in the ROM. This has been given the label ROMTOP. The B and C register pair are used to store the address of the lowest location in the ROM, here called ROMBOT. ROMTOP is assumed to contain the checksum value itself. The program then enters an addition loop in which every location except ROMTOP itself is added and stored in register B. When the contents of H and L have been decremented down to the value ROMBOT, the loop ends and H and L are again pointed to the checksum location ROMTOP. The value stored in this location is then compared with the total in register B and if they differ the program jumps to an error program called ROM-FAULT.

A similar program may be applied to any microprocessor by implementing similar mnemonic codes. Some commands may require extra instructions if no direct equivalent exists. The addresses used for ROMTOP and ROMBOT will also depend upon the microprocessor in use. Intel devices and Zilog Z80 locate ROM at the bottom of their memory maps because the program counter is set to zero at power on, while the Motorola 6802 and the MOS Technology 6502 locate ROM at the top of memory space.

If the checksum is stored in a separate ROM from the one being tested, then the routine will have to be modified to include the uppermost location in the addition loop and the checksum location will be altered to agree with its storage position.

9.1.2 RAM testing

The basic RAM test consists of writing a worst case bit pattern (55_{hex} or AA_{hex}) into every RAM location and reading them all back. If any location gives a result other than the pattern written to it, a fault is inferred and some form of error display given. In practice, these worst case bit patterns are not used and 00 is written to every location. RAM is assumed to power up in a random condition, which gives a probability of any location setting up to 55 or AA equal to setting to 00. Using 00 also serves the dual function of initialising RAM to zero before it is used by the system proper.

A general routine for testing RAM is as follows:

```
START RAMTEST
   RAMTEST: LD A, 00              ;clear Acc to zero
            LD HL, RAMBOT         ;point HL to bottom of RAM
            LD BC, RAMTOP         ;point BC to top of RAM
   RAMLOOP: LD(HL),A              ;write 00 to RAM location
            LD A,(HL)             ;read RAM location
            CP 00                 ;is result 00
            JP NZ,RAMFAULT        ;if not zero, jump to RAM fault routine
            INC HL                ;point to next RAM location
            XOR A,A               ;clear carry bit
```

```
            SBC HL,BC              ;subtract RAMTOP from contents of HL
            JP NZ, RAMLOOP         ;if HL not equal to RAMTOP, jump back
END RAMTEST
```

The program, written in Zilog Z80 mnemonics, initially sets the A register to zero. This is the value written to each RAM location. Register pair H and L are then pointed to the lowest location called RAMBOT. Similarly, register pair B and C are pointed to the highest expected location called RAMTOP. The loop routine, RAMLOOP, then writes zero to a location and reads that location back. If the result of the read operation is not zero, then a fault is inferred and the program jumps to a fault routine, RAMFAULT. If zero is read back, H and L are incremented to point to the next RAM location and the write-read cycle repeated. When eventually all the RAM locations have been tested, HL will point to location RAMTOP and the routine terminates.

The same routine can be used in a general purpose microcomputer to find the highest available RAM location. By setting HL to the lowest RAM location and omitting to place a value in registers B and C, the routine will continue to increment through memory until it encounters a ROM location which does not contain zero or it finds a memory space which is unoccupied in the system. As an example of this, consider that we have a system with the following memory map:

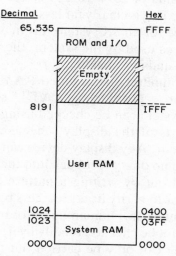

This memory map is typical of a small microcomputer based on the Motorola 6802 or MOS Technology 6502 microprocessors. ROM is located at the top of the memory space, up to location FFFF$_{hex}$ (65535$_{dec}$). The bottom end of memory contains RAM but is reserved for use by the system itself. In these locations will be stored often used system constants and the system stack. User memory is assumed to start at location 0400_{hex} and increases upwards to $1FFF_{hex}$. Thus the system contains 8K of RAM but only approximately 7K is available to the user.

A routine similar to the one given for testing RAM can be used to initialise RAM to zero and to determine its size. RAMBOT will be specified at 0400_{hex} and the routine will write and read zero to, and from, every user RAM location until the program encounters location 2000_{hex}. This location will be the next one above the top of RAM where no memory device exists in the system. A non-zero value will be read back and the routine will assume it has encountered the top of RAM. Using decimal values, the first location that defeats the write/read zero test will be location 8192. From this value will be subtracted the amount of system RAM at the bottom of memory to give a user size of 8192–1023, i.e. 7169 bytes free. Many micro-computers display this value on a VDU screen as a check on the size of system available. If now an additional 8K of RAM is added, the top will increase to address $2FFF_{hex}$ and the test routine will display a value of 16384–1023, i.e. 15361 bytes free.

This technique also provides a means of locating and indicating a faulty RAM. Assume for the moment that location $081E_{hex}(2079_{dec})$ is faulty. A non-zero read from this location will terminate the test routine and cause a value of 2079–1023, i.e. 1056 bytes free, to be displayed on the VDU. This is clearly far less than the usual displayed value of 7169 bytes free for an 8K system. From a knowledge of the size of the RAM devices used by the system, the user can work out which device is at fault and investigate it.

Many small microcomputers use a video RAM, each location of which maps directly to a position on a VDU screen. The RAM portion of the video system can be checked using the test program, but the remaining parts of the display scheme cannot be directly checked by the processor. Any display device can be thought of as a *Write Only* device with no direct feedback into the computer system. A test can be carried out by writing a pattern to the screen and requesting the operator to verify its correctness by depressing a key. Systems testing then involves a human operator interacting to prove those parts which cannot be checked readily by it.

Input/Output testing can only be carried out to a limited extent during power-up self-test checks because of the possibility of setting output lines to an unwanted and possibly dangerous state, and through the lack of feedback connections from external lines into the microcomputer. Input ports may be verified by connecting into them known stimulus patterns which are read by the computer and validated as being correct or incorrect. The stimulus pattern may be supplied from a test rig attached to the input under a request from the computer system or, in some cases, may be selected under computer control during a self-test sequence. Similarly, under certain conditions some output functions may be checked without affecting external circuits by disabling the final outputs through the use of a set of tri-state gates controlled from the computer. The output may be diverted into an input port and the bit pattern verified. These and other I/O related tests are covered in Chapter 10.

I/O testing usually involves either extra hardware in the system, if automatic testing is being carried out, which protects external circuits energised from the computer, or connecting test stimulators into the system upon demand from the computer. The latter approach is usually allied with diagnostic test routines which may be stored permanently in the system or loaded by test personnel and initiated by them.

9.2 DIAGNOSTIC TESTS

When a system fails to function in part and where the kernel can be shown to be functional from a free-run test, a user may invoke *Diagnostic* test routines to check out the rest of the computer. These tests may be stored in ROM which may either be permanently loaded in the system or plugged into an existing system ROM socket. Alternatively, tests may be loaded from some mass storage medium such as cassette tape or floppy disc. In many ways the first method is to be preferred, because a fault in the peripheral circuitry of the computer may itself prevent test programs from being loaded.

The open loop nature of the majority of I/O functions generally means that the user has to interact with the tests in order to verify the correct operation of circuits. Signature analysis test programs are an example which involve the user in connecting a signature analyser to the computer and verifying measured signatures against previously documented ones. In the absence of these routines and associated documentation, simple tests can be provided which exercise various parts of a system. Many computer systems connect to external devices via some form of connector which can be unplugged and a test box connected instead. The test box serves as a plant simulator and will usually contain switches to simulate inputs and indicators to display output bits. Where keyboards and VDUs form part of a system, they can serve as the test box themselves and also the dual role of enabling inputs and outputs to the system and providing a method of testing keyboard and display functions. A diagnostic test routine may, for example, display a message on the screen to press every key on the keyboard. Each character is then echoed onto the screen and the computer requests the user to verify them in turn by pressing a specified key. There are clearly flaws in this test method, because the verification key may be faulty. Whatever form the test box takes, the computer needs some defined stimulus pattern which it can verify as correct or incorrect to check input ports and some form of feedback from the tester to verify the correctness of any output port bit patterns it sends.

In larger computer installations several levels of diagnostic routines may be used. At the first level, the system is treated as functional sub-systems and each is exercised to check its correctness. A sub-system may be a floppy disc unit or a *Data Acquisition* unit. If a fault is indicated, then lower level tests may be run to check in greater detail, the failed unit. If a unit fails, clearly there will be a limit as to how far further testing can be taken. These detailed tests will be dependent on the particular hardware being tested.

9.2.1 Simple I/O testing

The majority of computer systems will contain at least one input and one output port which may be simple TTL or programmable devices. By connecting a set of switches to the input and a set of LEDs to the output port from a test box, both can be functionally tested. The following program illustrates the type of test that can be applied:

```
START SIMPLE I/O TEST
   I/OTEST: XOR  A              ;clear register A
            OUT  (OUTPORT),A    ; turn off all LEDs
   I/OLOOP: IN  A,(INPORT)      ;read in switches
            OUT  (OUTPORT),A    ;echo onto LEDs
            CP  81H             ;are MSB and LSB both set
            JP  NZ,I/OLOOP      ;if not then get new input pattern
END SIMPLE I/O TEST
```

The program assumes that a test box has been connected to an input port labelled INPORT and a set of LEDs to an output port labelled OUTPORT. Initially it turns off all LEDs and then reads in the switch states which are echoed back onto the LEDs. Each input and corresponding output bit can be tested for both logic states by flicking the switch on that particular bit. To provide an escape method, the routine checks if both the switch connected to the *Most Significant Bit* and the switch connected to the *Least Significant Bit* are both set at the same time. If they are, then the routine is exited.

The routine can be the first program in a diagnostic test ROM which is switched into replace the first system ROM and which can be initiated by resetting the system.

In certain instances the routine can be automated by connecting an output port to an input and running a routine which exercises every bit. The routine can check that the bit pattern sent is the same as that received and indicate a fault if a discrepancy occurs. Devices such as Zilog's PIO already have this type of feedback connection in each of its programmable I/O ports. Thus a port can be programmed to act as an output, while leaving its associated input port still active. Data can be written to the output port and read back into the computer from the same I/O address. While this verifies the computer I/O lines up to and including the pins on the PIO, it can only be implemented where the lines to peripheral systems can be protected. This may take the form of either disconnecting the peripheral device altogether or disabling the lines by putting a set of buffer gates into a tri-state condition. The enable/disable control for the buffers may be manually set or controlled from the processor. An example of this type of testing is given in Figure 74.

The PIO port is programmed as an output, but the 8-bit input port also remains active. Data can be latched to the output from the processor and read back into the CPU from the same address. The internal feedback connections between the output and input ports are contained within the PIO. The I/O lines from the PIO device are connected to external devices via a 74LS244 octal non-inverting

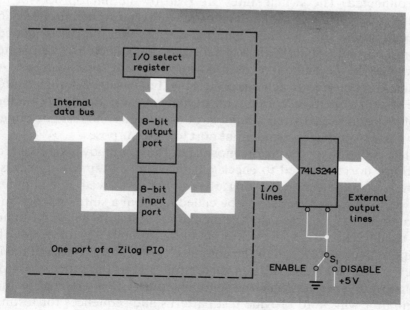

FIGURE 74   TESTING A PROGRAMMING I/O DEVICE

tri-state buffer with its output condition set by the state of switch S_1. When S_1 is grounded, the buffer is non-inverting; connecting S_1 to 5 volts puts its outputs into a high impedance state. S_1 may be a manually selectable link on a printed circuit board. Care has to be taken to ensure that the devices driven from the tri-state buffer do not see the high impedance state as a logic "1" level, which will be the case if standard TTL devices are in use. If all the output lines are double buffered through Darlington drivers, this situation does not arise.

9.2.2 Initiating Diagnostic tests

For the majority of small microprocessor based systems, ROM based *Diagnostic* routines are used to test the major components. The ROM may be a permanent fixture and jumped into either by switching it into circuit and resetting the computer or initiating an interrupt, with the ROM residing in the *Interrupt Service* routine memory space. In very small systems, a system ROM may be removed and the diagnostic device plugged into the vacant socket.

A diagnostic ROM will contain several short test routines. The user still has to provide some means of running each of them, or ALL routines are executed in a looped program which encompasses all of the devices. Either alternative is valid and used for system testing. Signature analysis test programs are often run in the latter mode, where each test is sequentially operated in a large loop. To implement the former technique, an input port is used into which a set of switches is connected. The switch states are read by the microprocessor to determine which test in the diagnostic ROM has to be initiated. To accommodate the switches, external devices may be disconnected from a port into which the test switches, are reconnected or a separate port may be permanently reserved for on-board slide switches. A resident set of slide switches is preferable, but has to be designed into the system from its inception. Unfortunately, the majority of systems test procedures are often treated as afterthoughts and test personnel have to resort to freeing an input port for test purposes.

The commoner types of diagnostic tests have already been covered and are normally used to check out the main system components. Where more extensive testing of peripheral, as well as computer devices, is needed, tests may be called up from a suite of programs contained on some mass storage medium. These tests will in the main be used to exercise devices external to the computer proper and will require that virtually all of the computer system itself is functional if they are to be loaded and run successfully. Thus the primary use of diagnostic programs which are stored on a system external to the computer will be to provide test patterns and sequences for testing peripheral systems.

Chapter 10

Testing Peripheral Related Functions

All the testing methods discussed in the previous chapters have been concerned with checking the functions of the computer itself. In most cases the computer only forms a part of a larger system which encompasses peripheral devices and systems. Provided the computer is functional, it can be used as a vehicle for providing test stimuli for the peripheral systems. By judicious design of the external circuits, many of them can be self-tested by the computer and provided with go/no go indicators. As systems grow in scale, the provision of such functional indicators considerably eases the burden of first line maintenance. To be effective, systems that incorporate this approach have to be "designed for testability", which means that all the extra hardware and software needed is provided from the system's inception stage onwards. The primary idea at this stage is to provide facilities to test complete functions but not devices, which would place an unecessary and uneconomic burden on the designer.

Allied to the need for greater testability in designs is the need to provide adequate test points. Testing of a rack mounted circuit board requires that an extension board is plugged into the card slot and the card then inserted into the extender. Provision should also be made for picking up power supply rails to power logic probes along with test points. Where possible the system documentation should include typical waveforms which can be checked using a conventional oscilloscope.

The interfaces that allow communication between a computer and a peripheral device or separate system may themselves be difficult to test. Typical examples are the IEEE-488 parallel interface bus, which uses transfer protocols which are unique to it, and the serial asynchronous transmission standard, RS232C. Normally, logic analysers are geared to capturing information from bus structured systems, which makes them of little practical use when testing a serial transmission line and they cannot handle the special mnemonic codes used for the IEEE-488 bus. Special purpose analysers are, however, available which cater for these requirements.

10.1 AN EXAMPLEOF SUB-SYSTEM FUNCTIONAL TESTING

In any computer based measurement scheme a commonly encountered sub-system will be an *Analogue to Digital Converter*, often mounted on a single printed circuit board. To read in electrical analogues of process parameters, measured by plant based transducers, the signals will be taken through an input multiplexing stage into an analogue to digital converter. To actuate proportional actuators such as proportional control valves, analogue output signals are needed which are furnished by using *Digital to Analogue Converters*. An analogue sub-system will contain both requirements, along with signal conditioning circuitry, on a single pcb.

Individually, analogue to digital converters may be tested by connecting to their inputs a known reference potential and running a program which displays the digital equivalent in some suitable manner. The converse function can be tested by running a suitable program which sends known bit patterns to the converter and measuring the resulting analogue output voltage with a DVM. The digital bit pattern may be cycled through every possible state to produce an output ramp waveform which can be displayed and measured on an oscilloscope.

An analogue sub-system may be self-tested by a computer and an indicator illuminated on the sub-system if it passes all the tests applied to it. To achieve this end, the circuit has to be designed for testability so that the requisite interconnections are catered for. A block diagram is given in Figure 75.

The analogue sub-system depicted in Figure 75 is representative of a large number of commercially available, single card, data acquisition systems. While the detail has been omitted, the feedback paths shown in the diagram are rarely implemented in practice; a direct consquence is that the computer by itself cannot be used to self-test the sub-system. The inclusion of a relatively small amount of extra hardware and power-up initialisation software can readily transform a non-testable system into one that can be self-tested. The ADC shown in Figure 75 can be checked at either end of its range by sending the appropriate address to the input multiplexer and converting either ground potential or some fixed stable reference voltage. Here it has been assumed that a unipolar converter has been used, but a bipolar device can be readily catered for by including a third, negative reference potential as an extra input to the multiplexer. The quality

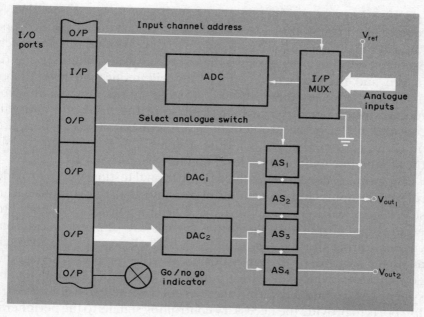

FIGURE 75   AN ANALOGUE SUB-SYSTEM DESIGNED FOR TESTABILITY

of modern converters is such that these measurements are sufficient to determine the state of the ADC on a go/no go basis.

Two DACs are shown in the diagram, and they may be tested by enabling analogue switches, which route their outputs back into the ADC. Each DAC may be cycled by applying known bit patterns to their inputs and converting their outputs, via the ADC, back into a digital quantity and comparing the pattern sent to that received. Each of the DACs may thus be functionally tested without adversely affecting any device to which they would normally be connected.

Provided that every major block in the system passes its tests, a go/no go indicator will be lit to indicate the acceptability of the analogue sub-system. Further indicators can be provided, to indicate "out of calibration", "DAC_1 faulty", "DAC_2 faulty", etc. The scheme lends itself to further refinement where input offset errors may be measured and allowed for in any subsequent actual measurements of plant variables; this process may be extended to the point where the sub-system is effectively calibrated, prior to any measurement being taken, by sampling the reference potentials and deciding offsets for every measurement. Taken to the extreme, this process will demand as many reference potentials, if the ADC is to be checked throughout its range, as possible inputs, but given the high quality of current ADC devices in practice will necessitate only a few reference points.

Self-testing any peripheral device or sub-system will almost always require the provision of extra hardware, which should be incorporated at the design stage. Again, taken to the extreme, this approach becomes self-defeating because the extra hardware will be prone to failure and the extra interconnections will reduce the overall reliability of the complete system. Testing at the functional level requires only a relatively small increase in the hardware and will considerably ease the problem of first line maintenance, where the objective is to minimise down time. Repairs can then be effected at the second line maintenance level on faulty circuit boards.

The primary concepts involved in functional testing are: to provide reference inputs and input multiplexers, so that the computer can select a known input, and verify it during a self-test sequence; to ensure that no possibly dangerous signals are applied to external circuits during the self-test sequence, by isolating outputs through analogue switches or tri-state gates; to provide comprehensive go/no go indicators, which allow test personnel to identify the failed sub-system easily. Implied within these criteria is the provision of extra I/O ports to provide the necessary controlling signals to the peripheral systems.

10.2 TESTING SERIAL DATA COMMUNICATION LINES

Information is usually conveyed between a computer and peripherals such as printers, VDUs and keyboards in a serial format. In its simplest form serial transmission needs only a pair of cables (the signal path and an earth return) which minimises the cost. Data within a computer is stored in a parallel form which has to be converted into a serial form for transmission. Conversely, data received over a serial line has to be reformatted into parallel form before being processed in a computer. The basis of these conversion processes is a shift register into which data is loaded in parallel and then clocked out serially, or into which data is clocked serially and when full is read out as a parallel word.

While there have been several serial communication standards, the most widely used is EIA RS232C. (The Electronic Industries Association is an American body which defines commercial standards. Their Recommended Standard, number 232, revision C, has become the accepted standard for serial transmission.) Originally this was defined to connect a terminal to a *Modem* (*Modulator/Demodulator*) to provide a universal method of connecting computer peripherals to a telephone line. It has, however, been adopted for connecting terminals into a computer system. RS232C defines a 25-way plug-socket in which every pin is allocated a signal whose meaning and direction are written into the standard. Pin 2, for example, is given the label XMIT DATA and is meant to convey information from the terminal to the Modem. RS232C also defines the electrical characteristics for the signals in terms of voltage levels, maximum transmission line length and speeds of data transfers. The following table gives the primary characteristics:

CHARACTERISTIC	RS232C
Maximum line length	100 feet
Maximum bits/second	2×10^4
Data "1"-marking	-1.5 to -36 volts
Data "0"-spacing	$+1.5$ to $+36$ volts

In practice the voltage associated with the two logic levels have been standardised on ± 12 volts.

The Modem should have a FEMALE connector and the terminal a MALE. The EIA has never defined the type of connector to be used with RS232C, but the industry has itself standardised on the 25-way D type.

RS232C is written in terms of a data terminal and a Modem; how then should a computer be classified in terms of the standard, as a Modem or as a data terminal? Most systems take the computer to be a piece of data communications equipment (a Modem) and fit a 25-way female connector so that terminals may be plugged directly into it. A problem arises when the computer is to be remotely accessed over a telephone link. It is illustrated in Figure 76.

Both the Modem and computer are fitted with female 25-way D connectors which, since the direction of data flow is defined in

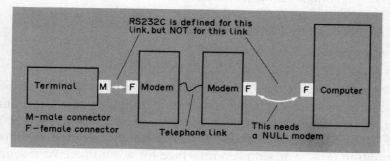

FIGURE 76   THE NULL MODEM PROBLEM

RS232C, means that both ends of the link will assume that they can send data over the same pin. Thus the Modem will attempt to send data to the computer over the same wire as the computer tries to send data to the Modem. Both will also attempt to receive data on the same line as each other. Communication between the two thus fails and to overcome the problem a NULL Modem is introduced. A NULL Modem simply consists of an interconnecting cable which has the same type of connector at either end; in this case both ends will have male connectors fitted but an alternative type of NULL Modem has female connectors. To allow data transfers, one end of the cable has signals switched between pins so that it appears to the opposing end as if it were the opposite type of equipment. Thus, the data send and data receive lines are swapped over along with their associated handshake control signals.

Twenty of the 25 pins in a connector are defined in RS232C, which caters for two transmission channels. The signal mnemonics are:

AB	Signal ground
CE	Ring Indicator (from Modem)
CD	Data Terminal Ready (DTR) (To Modem)
CC	Data Set Ready (DSR) (from Modem)
BA	Transmit Data (to Modem)
BB	Receive Data (from Modem)
DA	Transmitter timing (to Modem)
DB	Transmitter timing (from Modem)
DD	Receiver timing (from Modem)
CA	Request To Send (RTS) (to Modem)
CB	Clear to Send (CTS) (from Modem)
CF	Receive Line Signal Detector (from Modem)
CG	Signal Quality Detector (from Modem)
CH	Data Rate Selector (to Modem)
CI	Data Rate Selector (from Modem)
SBA	Secondary transmitted data
SBB	Secondary received data
SCA	Secondary Request To Send
SCB	Secondary Clear To Send
SCF	Secondary Receive Line Signal Detector

In practice it is rare to find a secondary channel in use. When used as a communications channel between a computer and a terminal, often, only the four handshake lines, Request To Send, Clear To Send, Data Terminal Ready and Data Set Ready, are used to control information flow. The simplest interface for a send/receive terminal uses only *Signal ground* (AB), *Transmit Data* (BA) and *Receive Data*

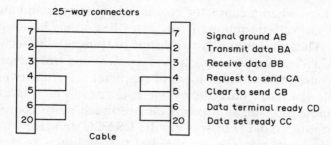

FIGURE 77   AUTO LOOP-BACK CONNECTIONS

(BB). Equipment at one end of the line might, however, expect to see *Clear To Send* at the correct signal level before it allows data to be sent. If the signal is not produced by the system on the opposite end of the cable, then it has to be hardwired into the connector to fool the device that it has received a handshake signal from the other end. The device will send out a handshake signal which is linked back into its own connector as a received handshake to its signal. Both connectors may be terminated in this way, which is referred to as the *Auto Loop-back* connection. Figure 77 illustrates this simplest of RS232C interconnections.

The auto loop-back connections may also be applied to a NULL Modem. It will differ from Figure 77 only by having pins 2 and 3 swapped over at one end of the cable and by having both connectors of the same type, i.e. both male or both female.

10.2.1 Serial transmission protocols

RS232C provides a common means of implementing a communications channel between a computer and peripherals. Before information can be passed between two such systems, however, both must be configured to send and accept data at the same rates, and in a format which both have been programmed to accept.

The most prevalent coding technique which allows alphanumeric and control codes to be passed between systems is the 7-bit ASCII (American Standard Code for Information Interchange) code. With few exceptions, this is universally used for the transmission of data over serial lines.

If we accept ASCII and implement RS232C there only remains the problem of deciding the order in which information is to be sent. This order is called the transmission protocol and governs not only the order in which the code for a character is sent but also the control bits needed to synchronise both ends of the line. The receiving end has to be informed when a character begins, where it ends and also needs some means of determining if any error has occurred during transmission. There are basically two forms of transmission technique in wide use called *Synchronous* and *Asynchronous* transmission. The former is used for sending blocks of characters at a time over a serial line while the latter only allows one character to be sent at a time. Synchronous transmission protocols are much more complex than asynchronous, and do not apply to the RS232C standard.

Asynchronous transmission requires that an initial *Start* bit is sent from the transmitter to the receiver to inform it that a character is about to be sent. When ASCII code is used, the 7-bit code for the character is then transmitted in reverse order. Thus the least significant bit is sent first and the most significant bit is sent last. Following these 7 bits, an error checking or *Parity* bit is then sent. The way in which the parity bit is implemented differs between systems. Parity may be ignored altogether and no bit sent, or even/odd parity may be used. If, say, even parity is in use, then the transmitter counts all the logic "1" bits sent in the ASCII codes character and if they add up to an even number, then parity is sent as a logic "0" level. If, however, the bits add up to an odd number, the parity bit is sent as a logic "1" level so that the total number of logic "1" bits received by the receiver is always an even number. Odd parity simply implements the converse of even parity. The single error checking bit can be used to detect all single-bit errors in transmission, but can be duped by multiple-bit errors. The receiver counts the number of logic "1" states it receives and flags an error if they disagree with the type of parity in use. After the parity bit, the line is allowed to idle for one, one and a half, or two clock periods before another character may be sent. These Stop bits are a carry-over from the days when most peripherals were electro-mechanical devices which could not respond instantly to another character and needed a dwell time between them.

The format for asynchronous data transmission is given in Figure 78.

Prior to any character being transmitted the line idles at +12 volts, which in RS232C represents the logic "0" state. If through some misfortune the line is severed or becomes disconnected at either end,

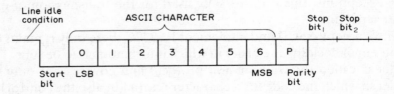

FIGURE 78   ASYNCHRONOUS TRANSMISSION PROTOCOL

its potential will fall to ground which can be detected at the receiver to indicate a line failure.

The transmitter clocks out information onto the line using its own clock, while the receiver clocks information into itself using a separate clock which conventionally runs sixteen times faster. The first negative transition on the line is taken to be the Start bit and the receiver counts 8 of its clock pulses and then samples the state of the line. This should place it in time half-way into the Start bit pulse. If the receiver samples a low-voltage state, then it takes the signal to be a valid Start pulse, otherwise it assumes a glitch and resets itself ready for a valid Start pulse edge.

If a valid Start pulse is detected, the receiver then counts a series of 16 pulses of its own clock which will cause it to sample the state of the line mid-way into each character bit and the parity bit. The sampled states are clocked into a shift register, which when full can be read, in parallel, into the computer system. If as an example the upper case alphabetic character A is being transmitted, this has a hexadecimal code of 41 in ASCII. Assume that even parity is being applied, then the parity bit will be sent as a logic "0" to ensure that the total number of logic "1" bits sent are even. The binary pattern for "A" is 100 0001 as a 7-bit code which when sent in reverse order with the parity bit included form the sequence 10000010.

The actual information on the serial data line is in a *Non-Return to Zero* (NRZ) format which means that between bits the line does not return to a logic "0" state. If, for example, two adjacent bits are both logic "1" states, then the line stays at that state for the duration of both pulse periods. Viewed on an oscilloscope, the transmitted character "A" with even parity will look as shown in Figure 79.

The transmit and receive clocks have to be within 2 per cent of each other if timing skew due to clock misalignment is to be avoided.

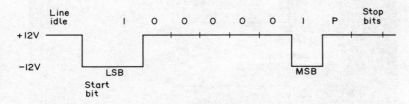

FIGURE 79   ASCII CODED "A" VIEWED ON A SERIAL LINE

10.2.2 Baud Rates

The rate at which characters are transmitted is quoted in bits/second or *Baud*. A *Baud Rate* refers to the total number of bits that are sent per character, including start, stop and parity bits. Typically a 7-bit ASCII coded character will have concatenated onto it one start bit, two stop bits and one parity bit, giving a total of 11 per character. Thus if the Baud rate is quoted at 110, then with 11 bits per character, 10 characters may be sent per second over the transmission channel. This low Baud rate applied to the early mechanical teletypes which were electromechanical devices. Later faster systems may employ rates up to say 9600 Baud, which again with 11 bits per character will enable up to 872 characters to be sent per second.

10.2.3 An asynchronous communications interface

To convert between the parallel format of information used within a computer system and the serial format used in an asynchronous transmission scheme, the essential components are a pair of shift registers. One register will handle data sent from the computer system while the second will receive data which will be read into it. The scheme can be visualised as shown in Figure 80.

Data is loaded into the *Transmit Register* from the CPU and then serially clocked out at the *Transmit Clock* rate. Received data is clocked into the *Receive Register* at a rate governed by the *Receive Clock* and when the register is full is read over the data bus by the CPU. This simple scheme, while illustrating the main functions, does lack certain vital elements. Extra logic has to be added to determine

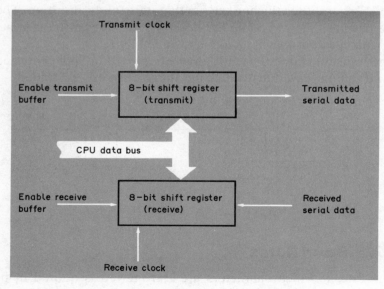

FIGURE 80 THE PARALLEL/SERIAL CONVERSION PROCESS

when the receive register is full and to check the received parity. The Start, Stop and Parity bits have to be stripped off, leaving only the character bits to be read by the CPU. The transmit section has to be provided with logic to generate the Start and Parity bits which have to be added to the character bits before a word is transmitted. All of these facilities are included in a single LSI device, the *Universal Asynchronous Receiver/Transmitter* (UART).

The UART was one of the earliest, mass produced *Large Scale Integrated* (LSI) circuits to be marketed. Early devices were programmed for such options as odd/even parity and the number of stop bits by strapping pins on the IC to supply potentials. Later devices could be programmed through software to set up the various options and were designed to be directly connected onto a CPU data bus. These devices also provided the major RS232C control signals, a typical example being Intel's 8251A USART. This device is fully programmable and can also handle several synchronous protocols such as SDLC. The 8251A is contained in a 28-pin DIL package and provides all the logic needed to generate Start, Stop and Parity bits. A block diagram of the device is given in Figure 81.

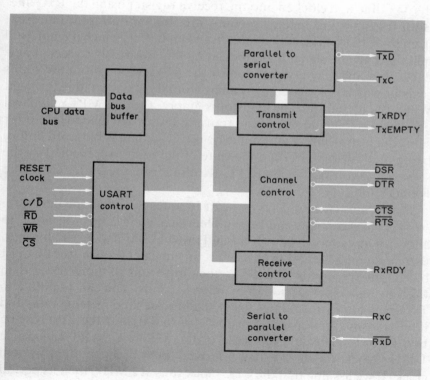

FIGURE 81   BLOCK DIAGRAM OF THE 8251A USART

Received data may be read from the device by pulling the $\overline{\text{C S}}$ (chip select) and $\overline{\text{R D}}$ (read) control lines low. Data to be transmitted is loaded into the USART by pulling both $\overline{\text{C S}}$ and $\overline{\text{WR}}$ (write) control lines low. To a programmer, the device appears as four registers, two of which may be written to and two that can be read from. The device is set up by writing to it with the $\text{C}/\overline{\text{D}}$ (control/data) line at a logic "1" level, while data is loaded into the device for transmission with the $\text{C}/\overline{\text{D}}$ line at logic "0". Similarly, if the device is read with the $\text{C}/\overline{\text{D}}$ line high, the status will be presented to the data bus.

On the serial transmission side of the device there are 11 signals, including separate clock lines for the transmit and receive sections. In many cases data will be sent at a different rate from received data. If we are interfacing to a VDU, then received data will appear at a relatively slow rate because it is initiated by manual entry from a keyboard, while transmitted data can be displayed at a much higher rate. Serial data sent from the USART emerges on the $\overline{\text{TxD}}$ line while data from some remote device, enters the USART on the $\overline{\text{RxD}}$ line. Data is shifted out of the transmit register using the clock applied to the TxC line and clocked into the receive register using the RxC line. Both transmit and receive sections are double buffered; data sent from the CPU is first loaded into a transmit buffer, which if empty will pull the TxRDY signal high, indicating that it is ready to accept new data. This line may also be read by the CPU as one bit in the status register which is accessed by reading from the device with $\text{C}/\overline{\text{D}}$ high. When the actual transmit register is empty, the TxEMPTY line will go high and the contents of the transmit buffer are loaded into it for subsequent transmission. Similarly, received data is buffered before being read by the CPU and the state of the buffer is reflected by the state of the control line RxRDY. Reading from the buffer clears the RxRDY line.

The receive clock must operate at 16 times the incoming data rate, so that character bits can be sampled near their centres in order to allow for any timing skew. If a Baud rate of 1200 is to be established, then, assuming 11 bits per character, the time per transmitted bit must be 833 microseconds. The RxC clock must run 16 times faster than this, giving a clock period of 52 microseconds. The clock applied to TxC must also run 16 times faster than the desired sending rate, and if the transmit section is also to operate at 1200 Baud then TxC can be tied to RxC. The internal logic of the 8251A USART requires a 1-MHz clock which is usually derived from the computer system clock. If we assume that the CPU clock runs at 1 MHz, then its period will be 1 microsecond; the common TxC and RxC clocks may be derived from this clock by dividing it by 52. This may be factored into

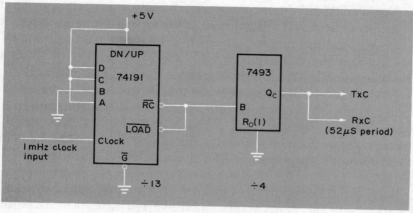

FIGURE 82 SERIAL CLOCK GENERATION

13 times 4 and the circuit shown in Figure 82 may be used to provide the serial clock signals.

The signal levels sent from and accepted by the USART are TTL compatible, i.e. ground and +5 volts nominally. The signal levels used by the RS232C interface are ±12 volts, which requires some form of level shifting circuitry before it can be connected to the USART. Integrated circuits are available to perform this function, commonly encountered devices being the MC1488 for converting TTL levels to ±12 volts and the MC1489 for converting ±12 volt signals to TTL levels. A typical interface, using these devices, for connecting the USART to the serial lines is shown in Figure 83.

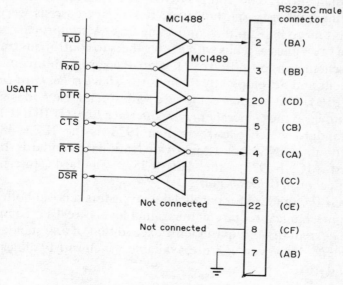

FIGURE 83 RS232C INTERFACE

To test an RS232C serial interface, an analyser is needed which can be set up to accept the Baud rate of the signals being analysed and which can capture and subsequently display the data preferably in ASCII. The analyser captures information by sensing the Start bit and then sampling the line to determine each bit as it is transmitted. A UART can be used to provide this function and convert each serially transmitted character into a parallel word for storage in RAM located within the analyser. After the RAM has been filled, a microprocessor located within the instrument can decode the stored bit patterns and display their ASCII equivalent characters on a CRT display.

Early serial line analysers captured information, after being set up for Baud rate and Parity sense, and converted the data into a parallel format for display on an oscilloscope screen. The Hewlett-Packard 1620A serial pattern recogniser provided information in this limited form. Later instruments such as the Hewlett-Packard 1640B serial data analyser displayed data in ASCII format on its screen. These later analysers could be used in a passive mode to simply capture and display information, or to simulate a terminal on an RS232C line and send, as well as receive, data from a computer serial interface.

10.3 MONITORING THE IEEE-488 INSTRUMENTATION BUS

The rate at which data can be transferred from one system to another can be dramatically increased if it is sent in a parallel format rather than serial. The complexity of the interface and the interconnecting cabling costs, however, rapidly increase as we include more and more lines in the connecting bus. A compromise scheme may use say 8 lines as a bus with a few others to control bus transfers. Such a scheme was proposed by Hewlett-Packard to interconnect test equipment and developed by them as the *Hewlett Packard Interface Bus* (HPIB). The bus has come into widespread use and is alternatively known as the *General Purpose Interface Bus* (GPIB); IEEE-488 1975 standard which was revised in 1978 as the IEEE-488 1978 standard; the ANSI bus (American National Standards Institute standard MCl·1-1975); the IEC 625-1 standard (International Electrotechnical Commission).

The GPIB standard for parallel data transfers sets out both electrical and mechanical requirements; signal levels are TTL compatible, but the current drive requirements exceed that of any standard TTL component. However, ICs are available which satisfy all aspects of the standard.

Data transfer rates can be up to 1 megabit per second and up to 15 major devices can be connected onto the bus at any one time. A major device in this context is a complete instrument such as a programmable DVM, a function generator or a *Floppy Disc Unit*. Within each major device, up to 32 secondary addresses can be referred to. A secondary address may relate to the d.c. voltmeter section of a *Digital Multimeter*, while another may relate to the ohmmeter.

Physically, devices connected on the bus must not be more than 4 metres apart, with a maximum transmission length of 20 metres, or twice the number of devices on the bus, quoted in metres, whichever is the least. Thus if only 3 devices are connected, the maximum transmission length reduces to 6 metres The IEEE-488 standard defines a 24-way connector, which is a combined plug and socket. These connectors are stackable and allow either a star or radial configuration of devices. It should be noted that the European standard IEC 625-1 specifies a 25-way connector, and to interconnect instruments designed to this particular standard onto an IEEE-488 bus adaptors are necessary.

The bus is intended to interconnect programmable instruments along with the bus controller, which manages all data transfers. Data transfers do not have to pass through the controller, but can be put onto the bus by one instrument and taken off it by several others under the supervision of the controller.

10.3.1 The GPIB structure

The GPIB uses 16 lines for communication; 8 for data transfers and a further 8 for control. The 16 lines may be viewed as three groups of lines: Data Input/Output (8 lines); Transfer Control lines (3 lines); Interface Management lines (5 lines). Each line is referred to by a mnemonic:

DI/O_1 to DI/O_8—Data Transfer lines

DAV   —Data Valid
NRFD —Not Ready For Data $\Big\}$ Transfer Control lines
NDAC—Not Data Accepted

EOI   —End Or Identify
IFC   —Interface Clear
SRQ   —Service Request $\Big\}$ Interface management lines
ATN   —Attention
REN   —Remote Enable

It is illustrated in Figure 84.

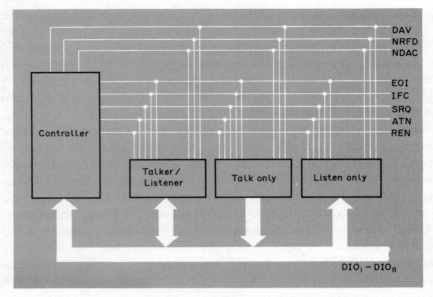

FIGURE 84   THE GPIB SYSTEM CONNECTIONS

The system can have only one controller, which determines the transfer of data from one instrument to one or more other instruments. Any instrument connected onto the bus can either be a *Talker* or a *Listener*. A talker can only place data onto the bus under the direction of the controller, while a listener can only receive data from the bus when directed by the controller. Any instrument may only be either a talker or a listener at any time, although they can be reprogrammed by the controller to change function, i.e. an instrument set up as a talker may be reprogrammed by the controller as a listener. The 8 bus lines DI/O_1 to DI/O_8 have to be used to send not only data from one instrument to another but also to convey address information. An instrument is set up as either a talker or a listener by the address bit pattern sent to it from the controller.

The overall rate of data transfer from a talker to listeners is controlled by the 3 handshake lines DAV, NRFD and NDAC. The interface management lines pass control information throughout the system, which may be initiated from either the controller or the talker placing data onto the bus.

EOI—End Or Identify

This line can be asserted true (low) by either the controller or a talker when the last byte of a message is being sent over the bus. It therefore indicates the end. It is also used in conjunction with the

ATN line to initiate a parallel poll sequence to decide whether any device on the bus requires servicing.

IFC—Interface Clear

This line can only be asserted true (low) by the controller and returns all devices on the bus to their idle state. It effectively acts as a general reset signal.

SRQ—Service Request

The SRQ line can be taken true (low) by any device on the bus that requires servicing by the controller; it effectively operates as an interrupt to the controller. It may, for example, be activated by a DVM when it has a new reading available. If SRQ is asserted, the controller begins a polling sequence using the EOI and ATN lines to determine which device caused the service request.

ATN—Attention

This line can only be activated by the system controller and when true (low) informs all devices on the bus that the information on the DI/O lines is address or control information. When the ATN line returns high, to indicate that the contents of the DI/O lines is data, only those devices previously activated as talkers or listeners can take part in transactions.

REN—Remote Enable

This line can only be activated by the system controller and must be held true (low) for any bus transaction to occur. If it is allowed to go false (high), all devices on the bus revert to local control. If a function generator has been manually set up from its front panel controls to produce a triangular waveform of a specified period and amplitude and the instrument has been reprogrammed over the bus as a sinewave generator, then if REN is taken high by the controller, its output will revert from a sinewave to the triangular waveform.

In some cases REN is permanently wired low to prevent any reversion to local control and SRQ may not be implemented; in such cases control is primarily exercised through the ATN and EOI lines.

All data transfers make use of the 3 transfer control lines DAV, NRFD and NDAC. The handshaking protocol used by these 3 lines ensures that valid data is not removed from the bus until all devices programmed to accept it have captured and stored it. All data transfers take place at a speed dictated by the slowest device on the

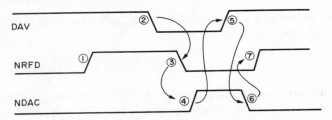

FIGURE 85   SIMPLIFIED GPIB HANDSHAKE PROCEDURE

bus. A simplified handshake timing diagram for these lines is given in Figure 85.

The DAV line is controlled by the talker, while the NRFD and NDAC lines are controlled by the activated listeners on the bus. DAV is used to indicate when valid data has been placed on the bus; NRFD indicates when listeners are ready or not to accept data and NDAC indicates when the listeners have received and accepted the data.

Assuming that the controller has set up a talker and programmed all the listeners, then all the listeners raise the NRFD line as shown by position 1 in Figure 85. The talker monitors the NRFD line and when it is raised high is aware that all listeners are ready to accept the data that the talker will place on the bus. The talker then places its data onto the bus and allows a short time interval to allow the new line states to stabilise. After this time the talker pulls the DAV line low to inform all the listeners that the data on the bus is valid. This state is shown by position 2 in Figure 85.

All listeners sense the DAV line going low, and pull the NRFD line low when it is ready to accept the data. After each listener has strobed the data into its internal buffers, it releases NDAC line to show that it has accepted the data. Only when the slowest listener has released the NDAC line can the sequence continue. This is indicated by position 4 in the diagram.

The talker senses that the NDAC line has been pulled high (false) and raises the DAV line to show that the data on the bus is no longer valid. Each listener in turn detects the DAV line going high and drops NDAC low to acknowledge the fact that data has been removed from the bus. This is indicated by position 6 in Figure 85. Each listener then raises NRFD high (false) to indicate that it is ready to receive the next byte of data sent down the bus.

The sequence is now complete, with all listeners waiting for the next data byte—this sequence is the essence of all data and address transfers over the bus and allows for signal propagation delays and listener processing time.

10.3.2   An example of a bus transfer on the GPIB

Let us assume that the word FREDDY has to be sent over the GPIB to a listener whose primary address is 05. We will assume that the controller has already sent the talk address to the sending instrument which will enable it to place data onto the bus. The bus transaction will be similar to that shown in Figure 86.

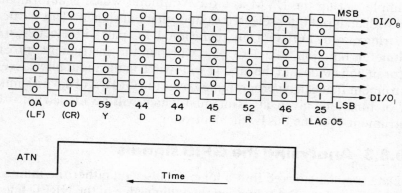

FIGURE 86   TRANSFERRING THE WORD FREDDY OVER THE GPIB

Characters are sent over the GPIB as ASCII coded characters when the ATN line is high, to indicate data is being transferred. In the example shown in Figure 86 it has been assumed that the word FREDDY is to be sent to one listener whose primary address is 05. When ATN is low, address information is passed over the bus; the first transfer shown sets up device 05 as a listener. The 3 most significant bits of an address transfer inform the specified device if it is to act as a talker or listener. In this case, the 3 bits are 001 which sets up device 05 as a listener. If the device is being set up as a talker, these 3 bits will be 010. Primary and secondary addresses are referred to as groups which are allocated mnemonic codes; the code for this particular transfer is LAG 05, which indicates that it belongs to the *Primary Listen Group*. The lower 5 bits of an address byte give the primary device address, in this case 05.

Each character in the word is then sent in ASCII format with the ATN control line high to indicate that data is being transferred. This particular word is delimited by sending a *Carriage Return* (CR) followed by a *Line Feed* (LF) at the end. During the transmission of the line feed character, the EOI line will be asserted low to indicate that this is the last byte of the transfer.

Every time that an address or a data character is sent over the bus, the handshaking protocol shown in Figure 85 is initiated. Thus for the example shown in Figure 86, there will be 9 such handshaking sequences.

An instrument such as a DVM will primarily function as a talker and send data back over the bus to listeners and/or the controller. To set up such an instrument, it will be addressed by the controller as a listener and then sent a string of control characters to program it to its required mode of operation. The string of ASCII characters may, for example, set up the DVM as a d.c. voltmeter which is autoranging and with an input filter selected to smooth input signals. This particular string may be FOROJ1 where the initial FO pair will select the d.c. voltmeter function, the RO pair will put the instrument into its autoranging mode and the final J1 will switch in the input filter. Every instrument that can be programmed over the GPIB will have a similar set of functions given in its manual which will be loaded into the instrument over the bus by the controller.

10.3.3 Analysing the GPIB signals

Data may be clocked into a logic analyser on either the falling or rising edges of the DAV line, on the falling edge of the NRFD line or on the rising edge of the NDAC line. Clocking data on the falling edge of either DAV or NRFD will cause data to be captured before it is transferred, while the rising edge of DAV or NDAC will only permit tracing of data after it has been accepted by all listeners.

There are several analysers which can be used to study GPIB transfers; the Hewlett-Packard model 1602A will allow up to 64 bus transactions to be stored and subsequently displayed in one of several formats (including ASCII) on a single line display. The Tektronix 7D01 logic analyser, which plugs into a Tektronix 7000 series oscilloscope mainframe, if fitted with a DF2 data formatter can be used to display GPIB mnemonics along with interface management control lines when asserted. A typical line would be:

ATN   LAG   18   REN   1011

which indicates that the Attention line has been asserted low (true), thus indicating that the bus contains address information. The LAG 18 indicates that the device with a primary address of 08 is being set up as a listener.

The final part of the display shows that the *Remote Enable* signal is true, thus placing all devices on the bus under the controller. The last 4 digits are meant to represent the states of 4 separate inputs to the analyser which may be used to monitor other signals in the system.

The 7D01 analyser with DF2 formatter can be used to display up to

17 bus transfers at any time on an oscilloscope screen, while up to 254 can be stored within the instrument and scrolled through on the screen display. This particular analyser can be switched into a timing mode of operation which makes it useful for studying timing problems on the bus such as incorrect handshake sequences.

References

General Digital Fault-finding

"Techniques of digital troubleshooting", Hewlett-Packard Application Note 163-1, July 1973.

"The IC troubleshooters", Hewlett-Packard Application Note 163-2.

M. Slater and B. Bronson, *Practical Microprocessors*, Hewlett-Packard, March 1979.

Logic Analysers

B. Farley, "Logic analyzers", *Digital Design*, January 1980.

Staff of Hewlett-Packard, "Logic analyzers", *Digital Design*, December 1978.

S. A. Rickman, "Selecting a logic state analyzer", *Digital Design*, June 1979.

"Mapping, a dynamic display of digital system operation", Hewlett-Packard Application Note 167-6.

"Engineering in the data domain calls for a new kind of digital instrument", Hewlett-Packard Application Note 167-4.

"The role of logic state analyzers in microprocessor based designs", Hewlett-Packard Application Note 167-13.

J. Brampton, "State analyzers move from lab to production area", *Electronic Design*, May 1982.

Signature Analysis

Hewlett-Packard Journal, October 1979.

Hewlett-Packard Journal, May 1977.

L. Palley, "E²PROMs bring flexibility to in-system signature analysis", *Electronic Product Design*, February 1983.

D. Peacock, "Signature analysis aids production testing", *New Electronics*, December 1979.

R. Beers, "Dual-mode signature analysis aids instrument troubleshooting", *EDN*, April 1982.

"An introduction to data-compression measurement for digital logic", Data brief number 301, Solatron Instrumentation Group.

Testing Peripheral Systems

"Using the 1620A for serial pattern recognition", Hewlett-Packard Application Note 167-10.

"Monitoring the IEEE-488 bus with the 1602A logic state analyzer", Hewlett-Packard Application Note 280-2.

210

Index

211